THE FOCAL EASY GUIDE TO
ADOBE® AUDITION™ 2.0

The Focal Easy Guide Series

Focal Easy Guides are the best choice to get you started with new software, whatever your level. Refreshingly simple, they do *not* attempt to cover everything, focusing solely on the essentials needed to get immediate results.

Ideal if you need to learn a new software package quickly, the Focal Easy Guides offer an effective, time-saving introduction to the key tools, not hundreds of pages of confusing reference material. The emphasis is on quickly getting to grips with the software in a practical and accessible way to achieve professional results.

Highly illustrated in color, explanations are short and to the point. Written by professionals in a user-friendly style, the guides assume some computer knowledge and an understanding of the general concepts in the area covered, ensuring they aren't patronizing!

Series editor: Rick Young (www.digitalproduction.net)

Director and Founding Member of the UK Final Cut User Group, Apple Solutions Expert and freelance television director/editor, Rick has worked for the BBC, Sky, ITN, CNBC and Reuters. Also a Final Cut Pro Consultant and author of the best-selling *The Easy Guide to Final Cut Pro*.

Titles in the series:

The Easy Guide to Final Cut Pro 3, Rick Young

The Focal Easy Guide to Final Cut Pro 4, Rick Young

The Focal Easy Guide to Final Cut Express, Rick Young

The Focal Easy Guide to Maya 5, Jason Patnode

The Focal Easy Guide to Discreet combustion 3, Gary M. Davis

The Focal Easy Guide to Premiere Pro, Tim Kolb

The Focal Easy Guide to Flash MX 2004, Birgitta Hosea

The Focal Easy Guide to DVD Studio Pro 2, Rick Young

The Focal Easy Guide to Discreet Combustion 3, Gary M. Davis

THE FOCAL EASY GUIDE TO
ADOBE® AUDITION™ 2.0

ANTONY BROWN

ELSEVIER

AMSTERDAM • BOSTON • HEIDELBERG • LONDON • NEW YORK • OXFORD
PARIS • SAN DIEGO • SAN FRANCISCO • SINGAPORE • SYDNEY • TOKYO
Focal Press is an imprint of Elsevier

Focal Press is an imprint of E
Linacre House, Jordan Hill, Oxford OX2 8DP, UK
30 Corporate Drive, Suite 400, Burlington, MA 01803, USA

First edition 2006

Copyright © 2006, Antony Brown. Published by Elsevier Ltd. All rights reserved.

The right of Antony Brown to be identified as the author of this work has been asserted in accordance with the Copyright, Designs and Patents Act 1988

No part of this publication may be reproduced, stored in a retrieval system or transmitted in any form or by any means electronic, mechanical, photocopying, recording or otherwise without the prior written permission of the publisher

Permission may be sought directly from Elsevier's Science & Technology Rights Department in Oxford, UK: phone (+44) (0) 1865 843830; fax (+44) (0) 1865 853333; email: permissions@elsevier.com. Alternatively you can submit your request online by visiting the Elsevier web site at http://elsevier.com/locate/permissions, and selecting *Obtaining permission to use Elsevier material*

Notice
No responsibility is assumed by the publisher for any injury and/or damage to persons or property as a matter of products liability, negligence or otherwise, or from any use or operation of any methods, products, instructions or ideas contained in the material herein. Because of rapid advances in the medical sciences, in particular, independent verification of diagnoses and drug dosages should be made

British Library Cataloguing in Publication Data
A catalogue record for this book is available from the British Library

Library of Congress Cataloging-in-Publication Data
A catalog record for this book is available from the Library of Congress

ISBN-13: 978-0-24-052018-6
ISBN-10: 0-24-052018-1

For information on all Architectural Press publications visit
our website at www.books.elsevier.com

Typeset by Charon Tec Ltd, Chennai, India
www.charontec.com

Printed and bound in the Canada

Contents

4   Editing (Edit View)   33

5   Multitrack View   57

Acknowledgements

I would like to thank the team at Focal Press, especially Catharine Steers for all her support and patience throughout this journey. Also, I wish to acknowledge Kevan O'Brien at Adobe; my wife, Zoë who has been a great source of encouragement to me; and my family: Jeffery, Sylvie and Kylie.

Preface

My first introduction to Adobe Audition was when a friend showed me how the software could remove a precise section of sound from within an audio file without affecting the surrounding areas. Since that time, I have used Audition on many feature films for cleaning up dialogues, removing dog barks, bird chatter, camera noises and many other sounds that no other software currently available can accomplish. Audition 2.0 is also a very efficient audio editing application to use when time is limited; it can breeze through many complex editing tasks in a short space of time.

The goal of this book is to get you up-to-speed as quickly as possible using all the tools that Adobe Audition 2.0 has to offer. Throughout this book there are diagrams accompanying the text to ensure that you understand all the procedures. Whether you are a complete novice or have a background in sound this book is for you. If you are using the Adobe Production Studio, which includes Premier Pro 2.0, After Effects 7, Encore 2.0; then getting familiar with Audition 2.0 will reinvent the way you work with audio within the Production Studio Collection. The integration between the Production Studio product lines makes the use of Audition easier than ever before.

Adobe Audition 2.0 is Adobe's main professional quality audio editing software application. It has a number of great functions including advanced restoration tools; fast editing functionality; a mastering rack with studio quality effects plug-ins; over 5000 loopology music content files to create complex music beds; low latency mixing engine with ASIO soundcard support throughout; and surround sound mixing capabilities. Video files can be imported into the Audition 2.0 timeline, and sound effects can be spotted with frame accuracy using both the thumbnail and main video monitor view. Audio files can be burnt directly to CD, and your files will be automatically converted to the correct format for this. You are also able to group normalise files to make sure all CD tracks are the same level.

I hope you enjoy reading this easy guide and that it will increase your knowledge and give you new skills for working with Adobe Audition 2.0.

Antony Brown

CHAPTER 1
SET UP

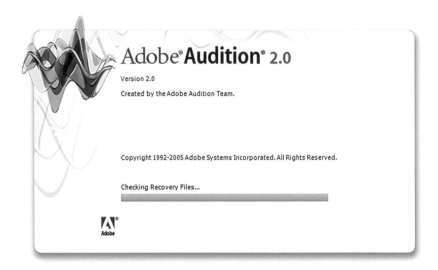

Loading Adobe Audition 2.0

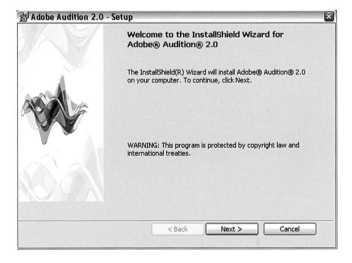

Installing Adobe Audition 2.0 on Windows XP is a straightforward procedure: simply insert the CD and follow the prompts. You will be prompted to associate file types that will enable your computer to launch Audition to play them instead of another application, such as Windows Media Player. If there is any specific file types you don't want Audition to open, then deselect them; otherwise press OK.

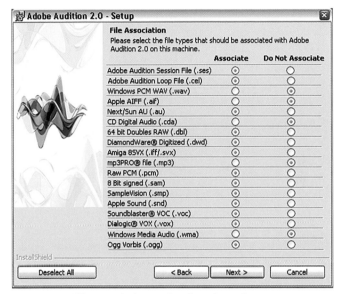

File association on loading the software

Note: If you have the Adobe Production Studio Premium edition software, which includes Audition 2.0, Premier Pro 2, After FX 7, Encore 2, Photoshop CS2 and Illustrator CS2, the one serial number supplied will load the whole collection. Or you can select individual components of the production studio collection (Audition 2.0 + Functional content) to load on your system.

Project Setup

Adobe Audition 2.0 will prompt you to set temporary folders on your drive to use. As Audition is dealing heavily with audio, it is best to use a separate drive if available for overall efficiency. When ready, go ahead and press OK and Audition 2 will open.

Setting Up the Device Hardware

Audition 2.0 now supports ASIO compatible sound cards for optimised low latency performance and audio scrubbing. It also works with any Windows-compatible soundcard, although latency performance will be reduced. Inputs and outputs for Edit View, Multitrack and Surround Encoder View are set from this window. Depending on the physical layout of your soundcard, different input and output port configurations can be set for each of the three main views within Audition.

To set up your audio hardware within Adobe Audition 2.0:

1 Choose>Edit>Audio Hardware Set up

2 Select the Edit View tab

3 Select appropriate ASIO soundcard from the Audio Drive drop down menu

4 Set the input and output ports

5 Repeat process for Multi-track and Surround Encoder Views.

Note: The 'Release ASIO Driver in Background' button releases the soundcard from Audition 2.0, so that other ASIO programs can use the soundcard device to play audio while Audition is open in the background. Highlighting Audition will re-allocate the soundcard back to Audition for playback – as long as the other application releases its ASIO driver too.

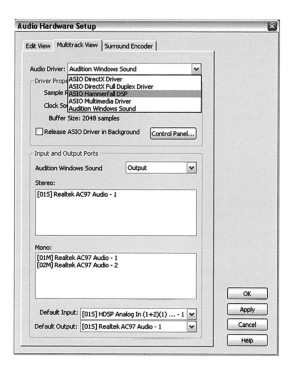

Latency

Latency in basic terms means the time it takes for sound to travel in and out of your computer via the soundcard. The standard Buffer Size setting of most onboard computer audio soundcards is around 2048, samples which result in a lag or delay in playback. An example of this would be talking into a microphone plugged into your computer soundcard and hearing a delay when monitoring

Latency: time it takes for sound to pass from Microphone – Computer soundcard – Speakers

the output; this is known as latency. Using Audition with an ASIO driver supported soundcard will allow you to directly monitor your input signal while recording with very low latency, sometimes known as 'zero latency'.

Preferences/key Commands

Preferences

Under the Edit>Preferences tab you have a number of options to fine-tune the way you use Audition 2.0, including:

- Display and colour scheme

- Memory use

- The amount of times you can use 'undo'

- Auto save – timed amount to save your project

- External controllers – assigning external mixer controllers i.e. Mackie Control

- Temporary audio folders – setting the Hard Drive space available.

Keyboard Shortcuts

The keyboard shortcut settings are another method for customising the way you work. If you are new to Audition but have knowledge of other software programs, keyboard shortcuts can be set-up within Audition to work as you like. To access keyboard shortcuts choose EDIT>Key commands.

New features in Adobe Audition 2.0

- ASIO driver support

- Direct to file recording

- Low-latency mixing engine

- Recordable parameter automation

- Hardware controller automation

- Audio mixing sends and inserts

- Real-time input monitoring

- Quick Punch

- Automatic delay compensation

- Effects chain on all channels, buses, and master

- Improved recording performance

- Unlimited tracks in Multi-track View

- Up to 96 live inputs and outputs

- Complete customizable workspaces

- Broader video format support

- Tighter integration with Adobe Premiere Pro and After Effects

- Improved user interface

- Custom workspaces

- Uncompressed 32-bit music loops

- Ready-to-use music beds

- Direct Adobe Bridge accessibility (note: the Bridge is *not* a part of Audition)

- XMP metadata support

- Audible scrubbing

- Analog-modeled multiband compressor

- Mastering rack in Edit View

- Enhanced frequency space editing

- Lasso tool

- New Spectral Controls, Spectral Pan and Phase display

- Save CD layouts

- Enhanced Broadcast wave support

- Improved performance for editing

- Ogg Vorbis format support

CHAPTER 2
THE INTERFACE

Introduction to the Audition 2.0 Interface

On launching Audition 2.0 you will be met with a brand new graphical interface, including new workspace panels that can be customised along with a revamped Mixer and Effects Mastering Rack.

Audition is made up of three main workspace areas: Multitrack View, Edit View and CD View, each view having its own panel layout and purpose. Each workspace is made up of separate panels that can be resized, docked or grouped to other panels within the workspace. Panels can also be floated separately.

In this chapter we will firstly look at an overview of all the different panels and interfaces and what each is used for.

1. Multitrack View

 The Multitrack enables you to record, edit and arrange multiple mono and stereo audio files into a musical session. The

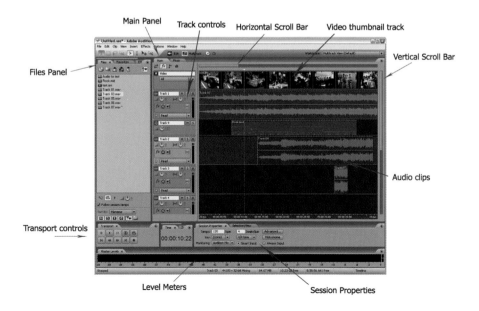

Multitrack View is made up of separate panels that can be resized and docked to other panels, customising your workflow. The Main panel is where multiple audio clips can be arranged along a timeline. A video clip can be imported, allowing you to accurately synchronise sound effects and music to the picture. There are controls to adjust volume, panning, add real-time effects and mix, all from within the Multitrack View.

Main Panel

Within the Multitrack View you have the Main panel showing audio and video tracks vertically, with video thumbnail preview and audio clips displayed horizontally across the panel. At the bottom of the panel the horizontal ruler or timeline shows hours, minutes and seconds. Different formats such as SMPTE time code and samples can be displayed by right clicking on the horizontal ruler>display time format and selecting the desired mode.

Main Tab

On the left side of the Main panel window there are four tabs showing overall settings for input/output, effects, sends and EQ within your Multitrack project.

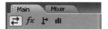

2. Edit View

The Edit View is Audition's main workspace for editing and mastering single mono and stereo audio files. Editing functions such as cut, paste, trim, normalise and mastering can all be done from within this view. Zooming functions allow you to select specific ranges of audio down to sample level for precise editing. The Edit View has additional spectral frequency displays that allow you to see and edit the audio in a Photoshop style way.

Edit View Main Panel

3. CD View

The CD View is where you assemble audio files into a track list ready to burn to a CD. A 'Group Waveform Normalise' function is also included to assure all tracks within the CD View have the same relative maximum volume level. The Files panel is located on the left showing a list of imported audio files. Located to the right is the Main panel showing a CD track list in chronological order. To the far right is the Track Properties area, showing file information, as well as move up/move down buttons to re-organise your files into the order in which you want them to play.

Video panel

Video files imported into the Multitrack View show a thumbnail display along the track, allowing you to locate scene changes along the timeline. The Main video

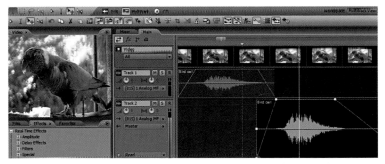

Docked Video Panel with Video Thumbnail Track in Timeline

Video Panel Undocked

Video Options

panel can be docked and resized within another panel or undocked in its own floating panel. Utilising both the thumbnail and main video panel you can position specific sound effects to match the picture at precise time code positions. You are able to customise the video panel for best fit, aspect ratio and quality.

Files Panel

The Files panel has the same function whether you are within the Edit, Multitrack or CD View. It displays all the audio, midi and video files that you have open. There are six buttons located at the top which include:

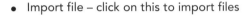

- Import file – click on this to import files

- Close file – removes any unwanted files

- Edit file – inserts the selected file into the main Edit View

- Insert into Multitrack session

- Insert into CD list

- Show Options button – located at the bottom of the panel is the audio preview function.

CU files panel tab

Effects Panel

The Effects Panel shows a complete list of included effects as well as any third party directX and VST effects you have loaded on your system. Both the Edit and the Multitrack Views have an Effects Panel list. The Edit View has process

FX, which can only be applied singly onto an audio clip; and also effects that can be applied either serially in the Mastering rack, or on their own. Some of the process effects have additional options available that the non-processed ones don't. In the Multitrack View, effects are applied to the whole track in real-time, so process effects can't be used here. In the panel, you can choose to group effects into categories as well as show just the available real-time plug-ins.

Favourites Panel

The Favourites Panel shows a list of frequently used effects including Vocal Remove, Fade-in, Fade-out and Repair Transient. Any favourite or frequently used process or effect can be added to the list to speed up your workflow.

To add to the favourites list go to:

1 Favourites>edit favourites

2 Select new>select desired function from the Audition effects list

3 Edit the settings>name

4 If you want to, assign a key command>press OK (this is not obligatory).

Tools Palette

 The Tool Palette shows a variety of tools that change the function of the cursor. Tools vary in function depending on whether you are in Edit or Multitrack mode. In Edit mode, you can only use the Marquee and Lasso tools if you are in the Spectral Frequency Display view. The four icons on the left are the Edit View tools; the four icons on the right are the Multitrack tools.

Transport/Time Format/Zoom Panel/Session Properties

These various panels show all the properties needed for your session, including time format, tempo, zooming functions and transport control. They are all resizable and dock-able.

| Transport | Time format | Zoom panel | Selection view | Session properties |

Audio Mixer Panel

The dedicated multi-channel low latency mixer is where individual adjustments are made to each track. Up to 16 different effects can be added per channel, along with a dedicated equaliser. The Freeze button allows you to conserve real-time processing power by locking the effects onto the track, although they can be unlocked again quickly if you want to change a setting. All mixer settings, including volume, panning, EQ, insert and send effects, can be automated.

Mastering Rack

A revamped Effects Rack for the Multitrack View and the mastering rack for the Edit view both allow you to add up to 16 chained effects. All effects parameters

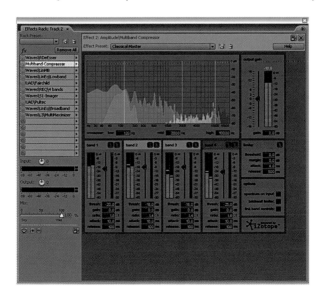

can be adjusted to get the best sounding result. Both mastering and effects rack have master input and output level controls along with pre and post fader options. The Mastering Rack has a preview button to let you hear exactly what your effects sound like before you hit the 'OK' key. The Multitrack effects rack runs in a real-time so parameters can be tweaked on the fly.

Spectral Views

The Spectral Frequency display gives you a Photoshop-style way of seeing the audio. Using colour to represent the frequencies and brightness to represent amplitude it is possible to see specific areas of sound that are not visible in the standard waveform view. New in Audition 2.0 are the Spectral Pan display and Spectral Phase display representing Pan and Phase information, as well as new Frequency and Phase Analysis tools. Intricate noise and tones can be individually selected using the Spectral Marquee and Lasso tools. Literally any discrete sound that you can see can be removed. This is ideal for removing unwanted sounds such as camera/background noise, dog barks, bird chatter, clicks and pops.

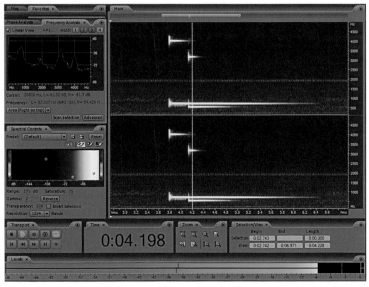

Spectral frequency display

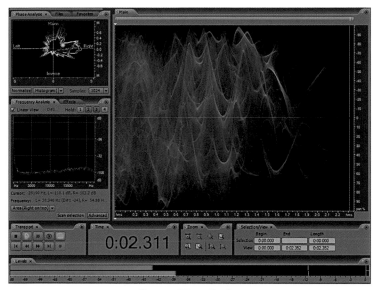

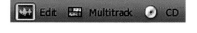

Spectral Pan display

Workspaces

1 To access the Edit, Multitrack or CD views of Audition 2.0 choose one of the three tabs located at the top of the screen.

2 To resize panels within a workspace, position your mouse between two panels. Click and drag to resize panels as desired.

3 When moving a panel to a new location, a blue shaded area will appear showing where the panel will be docked. Release mouse to dock the panel. To undock a panel hold down the CTRL key when moving or click on the arrow located top right and select undock.

4 To access a list of default workspace panels choose from the drop down menu under workspace, located top right of your screen or choose Windows>Workspace.

5 To save a new custom workspace, first arrange and resize your panels in the desired layout as described above. Then choose 'new workspace' from the workspace drop down menu and name and save your work-space>press OK. Your new workspace will appear in the drop down list with the other default workspaces.

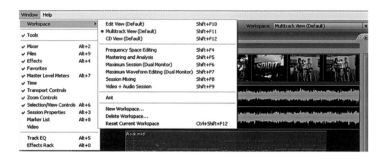

Note: If you lose a window panel at any time go to Window in the file menu, and select it from the drop down list. Now you can reposition it as you want.

CHAPTER 3
IMPORTING, RECORDING, PLAYBACK

Importing Files

In this chapter we will take a look at how to import all the major file formats into Audition 2.0, including audio files and extract audio from CD & video files. A full list of supported file types that Audition 2.0 can import are given in the supported drop down menu on the import dialogue window. I've listed in no particular order some commonly used ones below:

- Windows PCM (.wav)

- Apple AIFF (.aif)

- Broadcast wave file (.BWF)

- Windows media (.wma)

- Video files (.avi, mpeg, mpg, mp2 wma, mov)

- Audio from video (wav, aif, mpeg, mp3, wma)

- Audio from CD (.cda)

- Mp3, mp3 pro (.mp3)

- Audition Loop (.cel)

- Ogg format (.ogg)

Importing Audio Files in Audition 2.0

 There are several ways of initially opening the import dialogue box before you locate, preview and import files into Audition 2.0. Listed below are four different methods; they all have the same function, so use whichever one you feel most comfortable with.

1 Click once on the Import File button

2 Go to File>Import (only in Multitrack View)

3 Right mouse click in the Files panel (Multi-track or Edit View)>Import

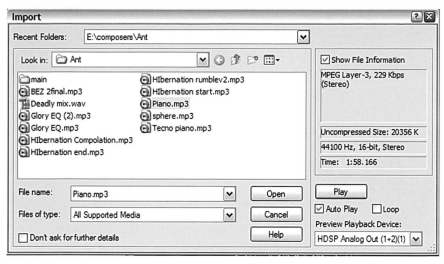

Import dialogue box

4   Double clicking anywhere in the files window area opens the import folder dialogue box

Once you have the Import dialogue box open, you can locate the desired audio file from your hard drive, then press open.

> **Note:** To preview the file before importing, highlight the file and press play or have the auto play box selected. If you can't hear any audio when previewing, make sure the preview playback device is set to the correct soundcard output.

Import/Extract Audio from CD

1   Insert a audio CD into your drive

2   Choose File>Extract audio from CD

3   Locate the CD track desired from the Extract Audio from CD dialogue box

4   Preview the track if desired

5   Press OK

Audition will then extract the audio from the CD, copying it into your project.

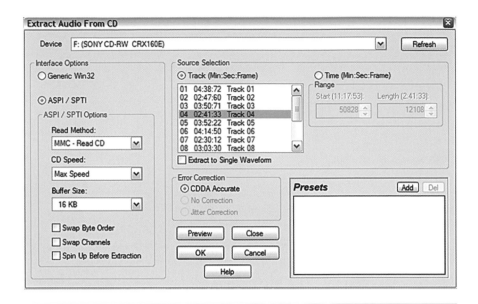

Note: you will need to save the file to a location on your hard drive.

Import Video Files

1 Locate a video file from the Import Dialogue box

2 Press open

Note: Both video and audio from the .avi file are imported separately into the files panel.

Import Audio only from Video Files

To import only the audio segment of a video file (avi, wma, mov or mpeg):

1 With Edit View selected choose Open Audio from video

2 Locate the video file that you want to extract the audio from

3 Press Open

Auditioning Files

Once files have been imported into the Files folder panel you can audition the audio files using the auto play preview function located at the bottom of it.

To audition files:

1 Highlight the file you wish to preview from within the Files folder

2 Press the play button located at the bottom of the Files panel

> **Note:** Turning on the Auto Play button will start playback as soon as the file is highlighted.

Recording Audio into Audition 2.0

In this section we will look at how to record audio into Audition.

> **Note:** You will need a microphone for this section.

You can record audio directly into both the Edit View and Multitrack View. The Edit View lets you record either a mono or stereo audio track, whereas the Multitrack View lets you record multiple tracks of audio at the same time. The amount of tracks you can record simultaneously in the Multitrack View depends on the amount of physical inputs your soundcard has, i.e. if your soundcard has 8 hardware inputs then you can record up to 8 individual tracks of audio into the Multitrack View at the same time. New monitoring modes in Audition 2.0 allow you to preview added effects and EQ to any track that you are recording, although the recording itself will not contain the effects you preview this way.

> **Note:** Monitoring EQ and effects in real time during recording will require an ASIO compatible soundcard for latency purposes.

Monitoring

The first stage in the recording process is to set the input signal level and monitoring mode. Audition doesn't directly control the input level of your soundcard, so you will need to adjust the level settings through the soundcard's own mixer software.

For this example we will look at monitoring setups for both onboard consumer based soundcards and the preferred Professional ASIO based soundcards.

Setting the Input Level

The input level is the volume level of the incoming signal. Ideally we want the input level to be as loud as possible without peaking the level over 0 dB (distortion) but at the same time, we don't want it to be too low or introduce unwanted noise. It's normal to leave what's known as 'headroom', so that sudden unexpected peaks don't overload. When using a microphone it is recommended to use an external hardware mixer, which has a dedicated microphone preamp for higher signal quality and input level adjustments.

Consumer Soundcard Input Level Adjustment

To set the input level of a consumer (onboard) soundcard you will need to open Windows Volume Control:

1   Choose start>control panel>sounds & audio device>audio

2   Set the playback and recording device if not already set

3   Choose Volume and set the record/input level for your input device

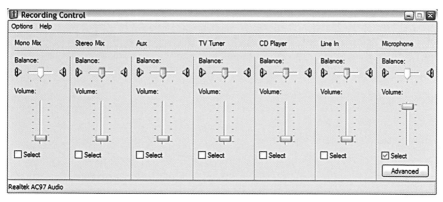

Setting the microphone input level on a consumer soundcard

Professional ASIO Driver Soundcard Input Level Adjustment

To set the input level of a professional ASIO compatible soundcard, choose the included mixer software supplied by the soundcard manufacturer and adjust

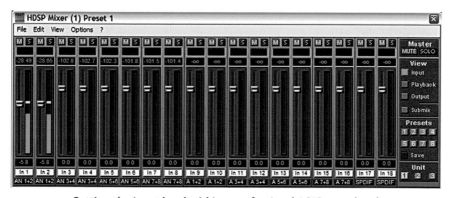

Setting the input level within a professional ASIO soundcard

input level. The benefit of an ASIO compatible soundcard is that it will have very low latency with little or no signal lag.

Setting the Monitoring Mode and Level

Monitoring mode

Monitoring modes

Audition 2.0 has two main monitoring modes: (1) External, and (2) Audition Mix. By setting the monitoring to 'External' you will be monitoring your input signal straight from the soundcard's output, decreasing latency (delay) – ideal for a consumer style soundcard. With a professional ASIO low latency soundcard, you can use the Audition Mix mode. This sends the input signal through Audition's mixer, so that you can monitor EQ and effects as you record, thus allowing you to preview the final mixed sound. Any EQ or effect present during recording is not printed to the track, and can be removed at a later stage simply by switching off the effect in the track's effects window.

Always Input

Once the selected Track Record Ready light has been switched on, Always Input will allow you to monitor your input signal at all times. Any underlying audio clip will not be heard when play or record is selected.

**Record ready
switched on**

Smart Input

Smart Input monitors the underlying audio clip in play mode (muting the input signal) until you press record to monitor the input signal, when it mutes the underlying audio clip – perfect for punch in recordings.

Monitoring Level

To set the monitoring level on a consumer soundcard, select the Windows volume control (as shown on consumer soundcard input/output level section), using the level controls in the Options>Properties>Playback controls section. For a professional soundcard running under the external mode, use its included mixer software for setting monitor level. If used under Audition Mix mode you select the Audition mixer for setting the monitoring level.

Metering

Right click on the meters to show input or out level metering options.

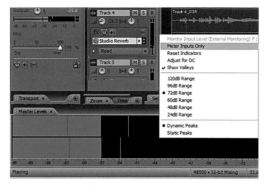

Metering showing input or output level

R To Record Audio into the Edit View

To record Audio into the Edit View:

1 Launch Audition if not already open

2 Choose the Edit tab to open the Edit View

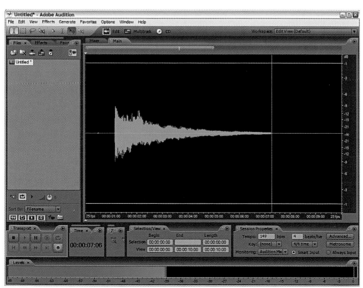

A new audio waveform appears within the Edit View as you press the record button

3   Choose file New (CTRL+N)

4   Choose new waveform properties including sample rate, mono or stereo and resolution

5   Press OK (A new untitled file will appear in the Files panel)

6   Simply press the **R** Record button on the transport bar to start recording

7   Press the stop or Record button to stop recording

8   Choose file>save as and name your file

R To Record Audio into the Multitrack View

To record multiple audio tracks into the Multitrack View:

1   Launch Audition if not already open

2   Choose the Multitrack tab to open the Multitrack View

3   Arm the tracks you wish to record onto by pressing the **R** Record button. Since this is the first track in a new session, you will be asked for a session file name, and a location to save it to at this point, otherwise you won't be able to record anything.

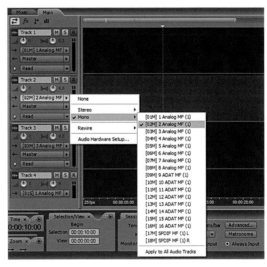

Setting the input for each track. Ready for recording

4   Set the soundcard input for each track located within the Main Panel View>input/output tab>inputs

5 Set record level and monitoring type (smart input or always on)

6 Insert any desired EQ and or effects located on the main panel (FX/EQ)

7 Press the record button on the transport bar to start recording

8 Press the stop or record button to stop recording

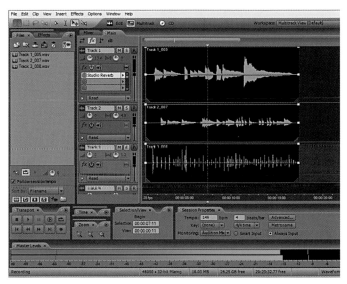

Shows three newly recorded tracks with reverb inserted on track 1

Punch In

If, at any stage within the recording process, you are not completely happy with your recorded performance, you can go back and punch-in a new take leaving the original audio clip intact.

To punch in a new audio take:

1 Select a range with the time selection tool on the appropriate track

2 Record enable the R track

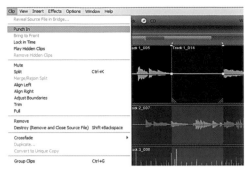

Punch In mode

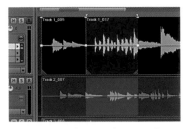

New recording within punch in range

3   Choose Clip>Punch In

4   Choose the Smart Input mode (As described in Monitoring section)

5   Reposition the cursor to allow a suitable preview time before the punch-in occurs

6   Press record and record your new take

Punch In mode automatically punches in and out exactly over the selected range. It leaves the original take untouched. Each punch-in records a new take, allowing you to select the one you want when you've finished recording.

Multiple takes can be made one over the other by right clicking the Record button and set to Loop while recording.

Multiple takes mode

CHAPTER 4
EDITING (EDIT VIEW)

Tools panel

Main Panel

Main View Tab

Effects Panel

Selected Range

Waveform Display

Transport panel

Level meter

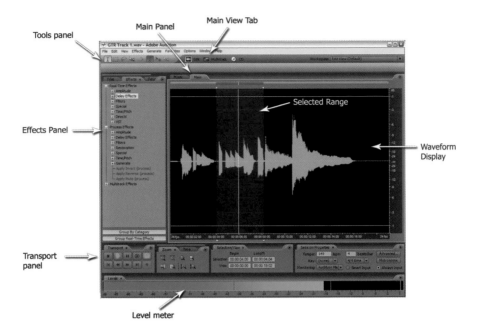

Introduction

The Edit View gives you a visual representation of either a mono or stereo audio file in wave form view. It enables you to edit and fine-tune the original file in a number of ways, either as a selected range or the whole audio file. The Edit mode operates in destructive mode whereby any editing will directly affect the original audio file which, in most cases, is what you want. Edit mode also comes with an accompanying Spectral View for seeing the audio in a Photoshop style way, for further editing capabilities which we will look at in more detail in Chapter 7 on Restoration. For the moment, bear in mind that any editing changes done to an audio file can be undone using the standard undo function right up to when you save the file. In this chapter we will learn how to:

- Zoom in and navigate around the Edit View workspace

- Select a range

- Cut, copy and paste

- Delete unwanted audio

- Add/remove silence

- Process FX

Destructive Editing

Edit View operates in a destructive editing mode where any edit, cut, paste or process you make will directly affect the original audio file when you save it. If your file is not finally saved, then it will revert back to its original state when it is closed.

Non-Destructive Editing

Within the Multitrack, View Audition 2.0 operates in a non-destructive mode. Any editing or processing is applied only to the way that the file (or sections of it) are played back, and the original is not altered in any way. The processing instructions are saved in a session file.

> Note: *Saving your work (Save copy as)* An important part of the editing process is saving your files. You have the option to override the original file (destructive editing) or save a copy (which leaves the original file unaltered and saves a new file with your adjustments).

Editing

How to import an audio file into Edit View:

1 Open the Edit View by clicking the main Edit tab

2 Import a mono or stereo audio file into the files panel using the import files button

3 Double click the imported audio file to bring it into the main waveform view; or with the audio file highlighted click the import file button.
Alternatively, pick it up with the mouse and simply drop it onto the Main panel

4 A waveform representation of the audio file will then be displayed, ready for editing

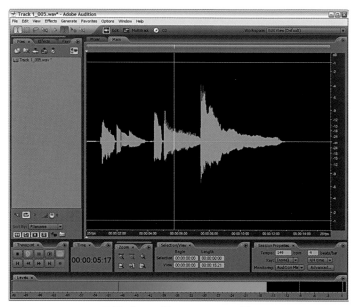

Edit view with mono wave file

Transport control

 The Transport panel covers all control movements.

Additional keys listed below:

- Press the space bar to start and stop playback

- Press enter to start and 0 to stop

- To rewind or fast forward use the transport control or left or right arrows

- Use the scrub tool located on the tools menu to hear the audio in forward or reverse whilst dragging the mouse over your waveform. Use the ALT key for
  Scrub tool different speeds while clicking the scrub tool, and the CTRL key for an alternative scrub effect.

Zooming and Navigating

Zooming

The Zooming tools allow you to zoom down to sample level enabling you to make precise edits and draw out troublesome clicks.

To Zoom in or out either:

1 Select the appropriate Zoom tool from the Zoom panel

2 Hover your mouse over the audio clip and use the wheel on your mouse to zoom in/out

3 Right mouse click and drag L/R on the horizontal zoom area at the bottom of the page

4 Right mouse click on the horizontal ruler>Zooming

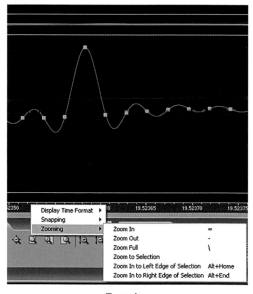

Zooming

Right mouse click and drag on horizontal timeline ruler to Zoom in

Moving

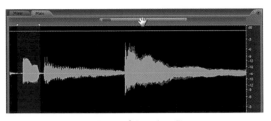

Horizontal Portion Bar

To move forward or backwards within the timeline area simply grab the blue Horizontal Portion Bar and move left or right. Zooming in/out can also be achieved by grabbing the front or end edges of the portion bar.

Snapping

As you may have noticed when selecting a range of audio, the time line snaps to different points along the timeline – this is because snapping is on by default (Edit>Snapping). This is a very handy way of quickly finding points along the timeline for precise editing. Snapping is defined by the display time format your project is set to, whether it be bars and beats, samples, or an SMPTE time code (frames per second). Additionally, Snap to Zero Crossings lets the ruler only snap to a point where the audio crossing is closest to 0 dB. This is ideal for stopping any annoying clicks or pops heard on cut points.

> **Note:** Try setting the snap to fine and see the difference.

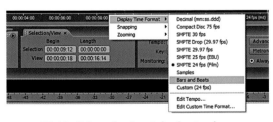

Right click on horizontal ruler to show Display time formats

To change Snapping settings, either choose Edit>Snapping or right click on the horizontal time ruler and choose Snapping. To change the display time format, choose View>Display Time Format or right click on the horizontal time ruler and choose Display Time Format.

Using the Range Tool

To select a range within the audio file:

1 Select the Time Selection tool from the tools menu

2 Use the Zoom and Snapping tools to fine tune your selection

3 Select a range within the audio clip by clicking and dragging the left mouse button across the audio

4 To adjust the in/out points of the range, right mouse click and drag the small yellow triangles in the corners toadjust the front and end points of the range

5 To select the whole audio clip simply double click anywhere within the clip area

Once you have a range selected, you then have many editing options available such as delete, move, silence, reverse, cut and paste and process effects, etc.

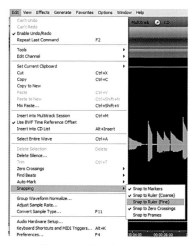

Choose Edit>Snapping

Choose Snapping settings by right clicking on horizontal ruler

Time selection tool

Right click on the horizontal ruler to set the Snapping

Range selected

Editing Functions Overview

In this section we will look at various editing functions within the Edit View, these include:

- Delete

- Cut

- Copy

- Paste

- Mix-paste

- Delete silence/add silence

- Mute

- Applying effects

To Delete Selection (Range)

To delete a selected range of audio:

1 Import audio clip into Edit view

2 Select a range of audio you wish to delete with the Time Selection Tool

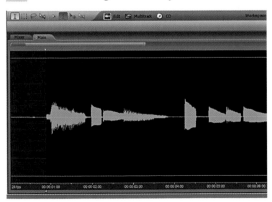

3 Use the Zoom and Snap tools to locate precise in and out edit points

4 Press the delete key

5 The image now updates and the selected area has been deleted

**Range selected at start of clip using
Time selection tool**

> **Note:** If you were to save this now (file>save) this would update the original file, otherwise known as Destructive Editing.

Cut/Copy and Paste within Edit View

Cut/Copy/Paste

1 Open audio file into the Edit View main workspace

2 Select a region of audio with the Selection Tool ⌶

3 Choose Edit>Cut or Edit>Copy (CTRL X) or (CTRL C)

4 Locate to new paste position along timeline

5 Choose Edit>Paste (CTRL V)

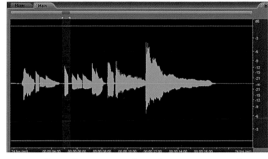

Selected range ready to be cut or copied

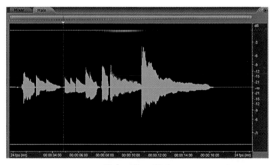

Selection Cut

Copy/Paste to New Location

There is also an option to copy/to new, which will let you copy or paste the selected audio range into a new window creating a new audio file.

Selection pasted further along timeline

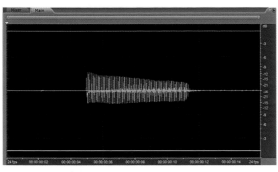

Selected range pasted to new location

To copy or paste a file to a new location:

1 Select the range with the selection tool 🖙

2 Choose Edit>Paste to new or Choose Edit>Copy to new

Mix Paste dialogue box

Mix Paste

The Mix Paste dialogue box lets you overlap the audio on top of the underlying audio. A Crossfade tick box option allows you to make a smooth transition or crossfade between the original, and the file you are going to paste. To change the length of the crossfade you will need to alter the crossfade Millisecond (MS) amount. L and R level controls allow you to adjust the relative levels of the audio to be mixed. You can also set the amount of times you want the selection to repeat, if you want it to, using the loop paste amount tick box.

Audio clip with Mix paste x3

Delete Silence

The Delete Silence function differs from the standard delete function as it allows you to delete silent areas throughout the whole audio clip automatically. Audition is able to find silence sections of audio by analysing the entire audio file. Settings to define silence and audio areas as well as markers, can be placed at each cut point to see exactly where silence has been removed.

To delete silence throughout the entire audio clip:

1 Open Audio file into main Edit View

2 Choose Edit>delete Silence

3 Define Silence and Audio amounts in milliseconds and dB (There's an optional tick box to show cut points in the marker list)

4 Select the Scan for Silence now then press OK

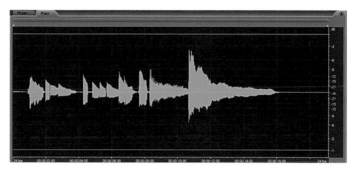

Original audio clip before silence

The 'after silence' clip shows that silent areas have been removed, leaving a trimmed audio clip.

Remember if you are not happy with the result, you can undo the process and redefine the silence and audio amount.

Delete silence dialogue box

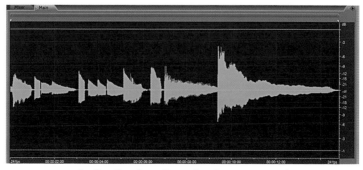

Audio clip after silence has been removed

Generate Silence

To add an amount of silence in seconds to an audio clip

Generate silence dialogue box

This increases the overall length of the audio file:

1 Locate a time-line position where you want to add empty space (Silence)

2 Choose Generate Silence (process)

3 Choose the amount of silence time

4 Press OK

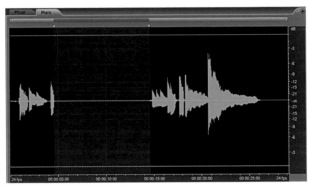

Silence added to file increasing overall length

Mute

The Mute function, unlike deleting or generating silence, does not affect the overall duration – it simply mutes a selected range.

To mute a selection of audio without affecting the length:

1 Open file into the Edit View

2 Select a range with the selection tool $\boxed{\text{I}}$

3 Choose Effects>Mute

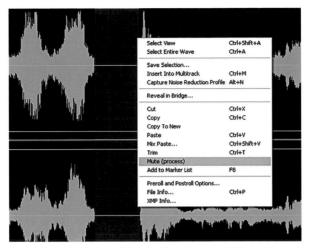

Mute process (without affecting overall length)

Using Effects in Edit View

In this Section we look at some of the effects included with Audition 2.0. When working within Edit View, effects are processed directly to the mono or stereo file that we are working with.

Applying Effects within Edit View

1 Open a file into the Edit View's main panel

2 Click on the Effects tab (on the left hand side you should now see a list of effects)

Within the Effects panel you will see a variety of categories. To apply an effect to your audio clip simply double click on an effect and it will open as seen in the Amplify example – see below. As these effects are applied directly to the file, you will need to preview them first to hear the resulting sound. At the bottom left there is a preview play button allowing you to do this before you press OK, which apply the effect to your file.

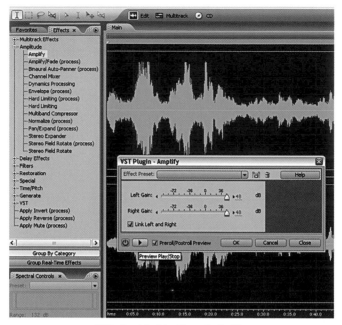

Effect added to the whole file

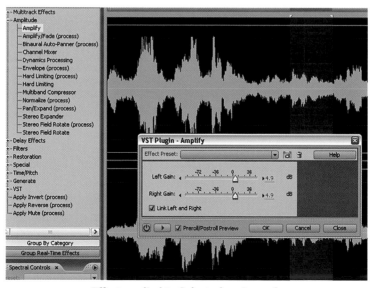

Effect applied to Selected region only

Note: By default the effect will be applied to the whole clip unless a range is highlighted *before you click on an effect.*

Effects Plug-In Overview

Adobe Audition 2.0 has a great range of included effects that can be used to:

- Fix audio problems
- Enhance
- Adjust pitch and speed
- Totally change or alter the sound of your audio

Let's take a quick tour of some of the included effects and their usages.

Normalize

The Normalize function enables you to alter the level of the loudest part of a file to any level you select. You can use it to turn up the overall level of an audio clip to its loudest point before clipping (distorting or going above 0 db) – the next diagram shows the Normalize effect plug-in being used for this purpose. The waveform behind shows half of the file without any normalizing applied (left) and the resulting normalized effected region (right). The Clip has been amplified equally to − .3 below just below 0 db. The vertical db level meter on the right shows the loudest peak waveform rising to just below 0 db.

Note: There is also a Group Waveform Normalize function that lets you normalize a bunch of audio files together (Edit>Group Waveform Normalize). See CD view for more details.

Hard Limiting

The Hard Limiter allows you to greatly increase the overall volume by increasing the input signal and then limiting the overall output level. The following diagram shows the resulting effect on the right with the level meters reaching the Max Amplitude setting. In this case − 3 dB.

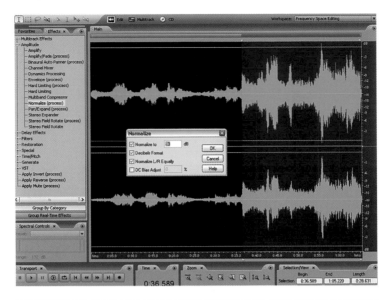

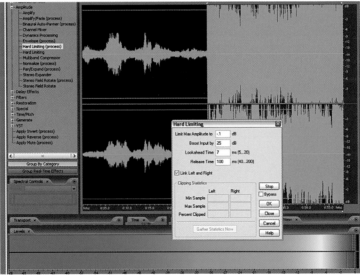

Delay FX

The Delay Effects category has a number of acoustic space effects creating a Grand Canyon style echo. Below you can see the list of Delay Effects including Studio Reverb.

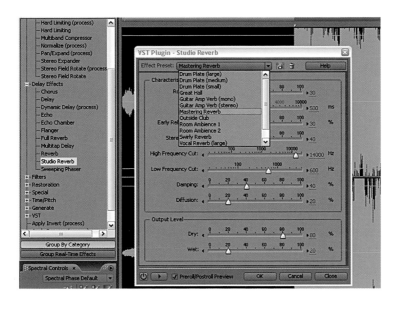

Amplitude Category: Multi-band Compressor

The Multi-band Compressor is a mastering tool that lets you compress or squash four-frequency bands independently (bass, low-mid, high-mid and high). It is possible to move the frequency crossover points and threshold levels and to increase or decrease the four band volume levels independently, as well as link all channels together. Also, the Multi-band Compressor is able to boost the overall master volume level to create a polished punchy mixed audio track.

Filters

The Filter Category has precision equalisers for fixing problem frequencies and enhancing the tonal quality of your audio. The Graphic Equaliser or tone control is a widely used effect, it is used on everything from a car stereo to an Apple ipod. Audition comes with a variety of equalisers under the Filter Section including this 30 band EQ.

Looking at the figure below, you can see that on the left the Graphic EQ has up to 30 frequency bands. This allows you to enhance, boost and decrease single frequencies. It is ideal for removing harsh unwanted frequencies, whereas the Dynamic EQ given below lets you adjust the amount of EQ over time. This will allow you to create a filter style sweep effect over time. The presets are a good starting point to build upon.

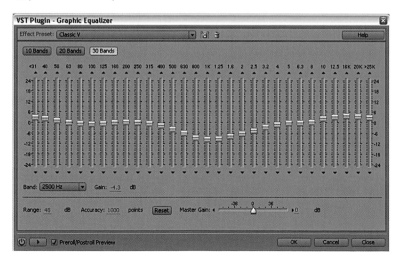

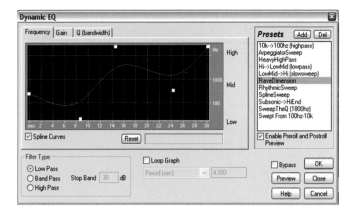

The other effects in the filter list – like the Scientific Filter – are for specifically getting rid of unwanted sound like Subsonic Rumble.

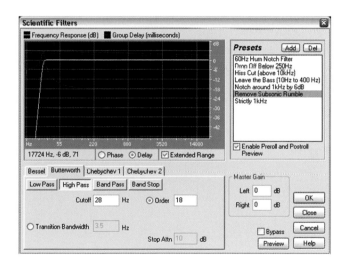

Converting Sample Types

Audition allows you to convert files to other formats, including sample bit rates and file type formats.

To check the file format:

1   Right click on the audio file>file info

2 It will tell you the file type, i.e. 48000 Hz 16bit, Stereo

3 Before you convert your files you may want to hear the result before processing

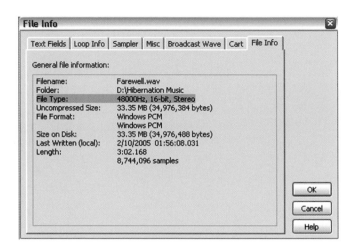

To convert the audio sample format:

1 Choose Edit>Convert Sample Type (F11)

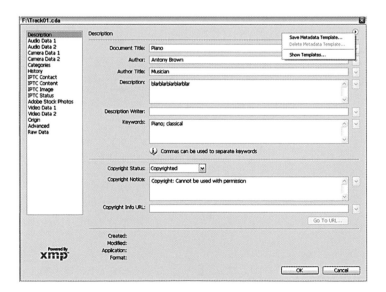

2 Select a sample rate from the list on the left, i.e. 44100

3 Select quality from low to high

4 Select mono if you want to convert your stereo file to mono

5 Set resolution

> **Note:** A standard CD quality file is 44.1 k/16-bit stereo and for film/TV choose 48 k/24-bit. If you intend on sending your files to be professionally mastered, then choose the highest quality 32-bit uncompressed (usually PCM wav) format for cross-platform integration.

Time-stretch/Pitch Correction

The Stretch Plug-in lets you change the tempo (speed) of your audio and can also be used to adjust the pitch or musical scale.

> **Note:** You may find that you only need to do one of the above.

Time stretch

Go to the Effects list and choose the Stretch effect in the Time/pitch category (Edit View only). To change the speed of your audio use the Stretch % slider to adjust to your desired speed. The quality of the stretch is determined by the precision settings (high quality will take longer to process but will give the best result). Also, clicking on the Constant Vowels tick box will help to keep vocal stretching relatively unnoticed.

Pitch Corrector

The pitch corrector offers an automatic way to fix problem vocal tracks that need the occasional note adjusted or

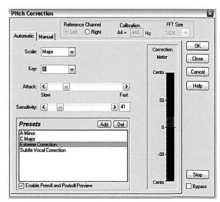

53

put back in-tune. To adjust the pitch, set the Transpose drop down menu to the desired setting and then set transpose to none if no pitch change is wanted. If you know the key and scale of your track, select them from the list and preview the result. If you have no idea what scale and key you're in, set the scale to chromatic; the key will now be set to Automatic. Set the preset to subtle vocal correction and then preview the result. Try experimenting with the Extreme Correction settings to get some weird sounding vocals.

Pitch Bender

The Pitch Bender can help you achieve the effects you may have made on your old record player when changing the speed from, say, 33 to 45. The speed would have gradually increased or slowed down – this is exactly the effect you can achieve using the Pitch Bender.

1 Draw in the speed curve points starting from left to right

2 The 'Zero Ends' button sets your start and end point back to where you started (if desired)

Adding XMP Metadata to Audio Clips

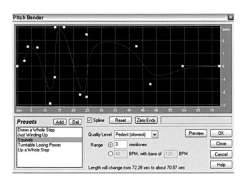

Adobe Audition lets you embed descriptive content information into the actual audio file. This Extensible Metadata Platform (XMP) allows specific information to be stamped onto the file, so that any XMP supported software can search, read, edit and share information across a database.

To add XMP metadata info to a file:

1 Select a file in the Files Pane under Edit View and choose File>XMP info

2 Select a category and type in the metadata information you want to add

3 Press OK

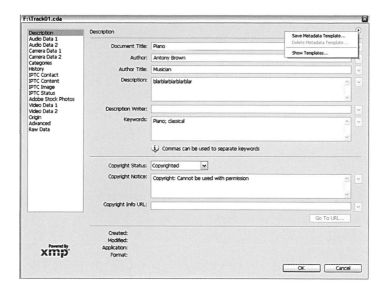

Note: you can also save metadata templates>located from top dropdown menu.

CHAPTER 5
MULTITRACK VIEW

Introduction

The Multitrack View is where you combine multiple audio clips to create a music or sound design work. Video can be placed on its own track, and sound spotted to exact frames along the timeline. Settings for EQ, volume and real-time effects can be accessed, and adjustment can be made to individual tracks.

In this Chapter we will look at the overall functions of the Multitrack View including:

- How to open and save sessions

- How to arrange clips on the timeline

- Editing functions

- Adding effects

- Applying cross-fades

- Automation procedures

Working with Sessions

Audition saves your overall project settings in a session (.ses) file. These are relatively small files that hold all the information about your Multitrack layout: such as where the edit and cut points are along the timeline, what effects and how many tracks are in use and, importantly, where your audio is located.

To create a new session:

1 Go to>file>new session

2 Select a sample rate, e.g. 48 khz

3 Press OK

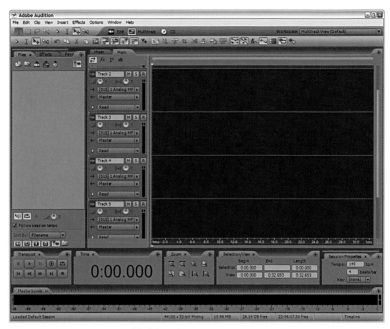

Multitrack default session

Saving a Default Session

To save your own default Multitrack session

This will open exactly the same way every time you open a new session:

1 Set up Audition exactly how you want the Multitrack panels to look, including track heights, view, FX inserts and outputs busses

2 Choose File>Default session>Set current session as Default Session

To open an existing project:

1 Go to File>Open (The Open Sessions panel opens with your Recent Folders opened at the top)

2 To locate sessions throughout your hard drive choose the Look in drop down menu. Highlighting a .ses file will display an information box showing particular file type information.

3 Once you have located the appropriate File type (*.ses) press Open

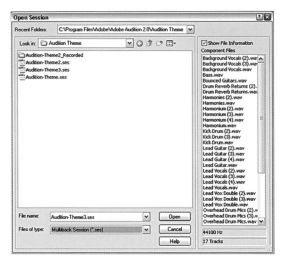

To save a session:

1 Go to File

2 Choose Save Session

> **Note:** This will save over any previous version with the same name.

Saving All

The Save All command saves changes to all the files open in your session, and the .ses file itself.

Saving a New Session

To save a new session under a new name keeping any previous versions:

1 Choose File>Save Session As

2 Choose which folder you want to save your project in

3 Click Options before saving if you want to convert the sample rate of your files

4 Press Save

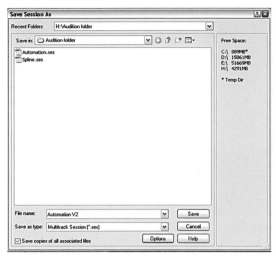

Save session As dialogue box

Note: For general housekeeping purposes it is recommended to tick the Save copies of all associated files. This allows all of the audio files related to your session to be neatly stored in the one folder, which is ideal for backing up your project.

Display Time Format

Once you have imported your audio clips into the Multitrack View you will want to start arranging them in some kind of order. To begin with, set the Display Time Format to the desired setting by right clicking

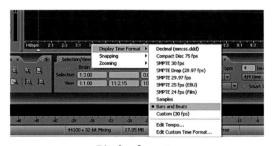

Display formats

on the horizontal time display. If you are working with musical loops select Bars and Beats; if working to film picture select the SMPTE fps (frames per second).

Importing Files into the Multitrack View

Similar to importing files into the Edit View choose file>import. Imported files are displayed within the files panel of the Multitrack View. When dragging files from

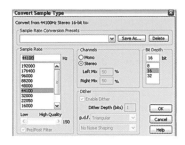

the files panel into the main panel you may be asked to convert the sample type if the file format of the audio file is different to the overall Multitrack settings. Set Bit Depth and Channel then press OK.

Snapping

Snapping settings

Snapping assists in aligning your audio clips along the timeline. When moving audio clips along the vertical timeline with snapping turned on clips will snap (line up) to markers, clips, loop endpoints, frames and multiple snap points can be selected. To set snapping right click on the horizontal ruler and select Snapping options.

Editing Procedures Overview

Listed below are a number of everyday editing procedures for Audition 2.0 including:

- Arranging and Moving Clips

- Copy and Paste

- Trimming

- Slipping

- Splitting

- Deleting

To move an audio clip:

1 With the Move Tool select the audio clip with a left click and move to new location

2 With the Hybrid Tool right click the audio clip and move.

To make a copy of an audio clip:

To copy a clip up/down the timeline

1   With the Move Tool  selected, right click and move the clip to new location

2   As you let go of the mouse a drop down menu appears

3   Select one of the following:

• Copy Reference Here: copies the original file; any processing done will affect the first clip as well as the copied audio clip

• Copy Unique Here: makes a new clip which can be altered and processed without affecting the original audio

• Move Clip Here: simply moves the audio clip to the new location without making a copy

Copy options

To copy and paste an audio clip:

1   Select the clips you want to copy/paste (Hold down CTRL to add more clips)

2   Choose Edit>Copy (CTRL+C)

3   Move to new timeline point>edit>paste (CTRL+V)

Trimming

To shorten the front or end of an audio clip otherwise known as trimming:

1   Place any currently selected tool at the edge of the start or end of a clip (the tool changes into a Trim tool ⊣|⊢)

2   Drag the edge of the clip to shorten or lengthen it

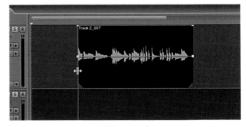

Trim tool

Note: You can only lengthen the clip to its total original length; to lengthen it further than its original length use the time stretch/rate stretch tool.

To split an audio clip (cut):

1 With the Selection Tool [I] locate to the point you want to cut

2 Choose Clip>Split (CTRL+K)

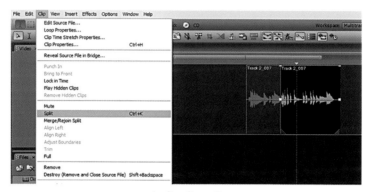

Audio clip spit

To trim the clip to a specific range:

1 Select a range with the Selection or Hybrid tool

2 Go to clip>trim, either by using the file menu, or right-clicking on your selection and selecting Trim from the options

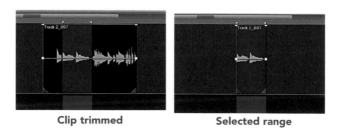

Clip trimmed   **Selected range**

To delete a specific range within a clip:

1 Select a specific range with the Selection or Hybrid tool

2 Press delete

Delete range within audio clip

To adjust the clip boundaries:

1 Select a range with the Selection or Hybrid Tool

2 Go to Clip>Adjust boundaries

To slip the internal contents of an audio clip:

If the audio clip's start and end points are in the right location but the internal audio is playing the wrong section you may want to Slip (move) the internal contents.

Internal content clip

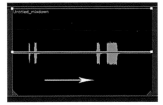

Internal content Slipped to the right within clip

1 Select the Move Tool

2 With the ALT key pressed>right click the audio clip and Slip (move) the internal audio

To rejoin split clip:

1 Select the clips you want to rejoin back together

2 Go to Clip>Merge/rejoin

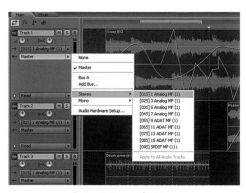

Output Bus selection

To set the inputs/outputs:

1 Highlight either the input or output button in the track control panel

2 On the selected track output button click the drop down menu and set the output to the desired soundcard destination

Note: You can send the track directly to the Master Bus, to any FX Bus you may have added, or to a specific output from your soundcard directly to your monitoring system.

Adjusting Volumes, Panning

Audition's main view has been updated in Version 2, making adjusting Volume, Panning, adding Insert effects and Sends, a whole lot easier and fluid.

Located at the top of the Main panel are four buttons

- Inputs/outputs

- Effects

- Sends

- EQ

Highlighting any button changes all the Track Settings accordingly.

The track settings are located below the buttons with track numbers, e.g. Track 1, Mute, Solo controls, Record, along with Volume and Pan knobs.

To adjust the Volume and Pan of a whole track from the Main panel:

1 With your mouse hovered over the volume or pan knob

2 Click and hold turning to the left or right to increase or decrease level

3 To reset Volume and Pan back to 0 hold down the ALT key and click

Adding Real-time FX from Multitrack View

Highlight the Effects tab to display the FX view. Below the Volume and Pan knobs you have the FX insert boxes which allow you to insert up to 16 effects per track.

There is an Effects Power Button to turn on/off the whole FX list and a Pre/post Fader Button that is used to insert FX before Sends and EQ (pre fader), or after Sends and EQ (post fader). Also, you can turn off individual effects by clicking on the Power button

beside them. The Freeze button allows you to freeze or process the whole track with the effects you have inserted. This will help to free up your computers processing power for other tasks. If you need to make an adjustment, simply unfreeze the track, make adjustments and freeze again.

To add real-time effects:

1 Click on the ▶ located on the right-hand side of the FX insert window to show all FX list

2 Insert an effect

3   Or drag and drop an Effect from the main Effects Panel onto the track itself in Main View

> **Note:** Effects can be changed in real-time as your session is playing

Setting the EQ

Highlighting the EQ tab ⇄ *fx* ⊩ ▥ opens a three band Equaliser that will work over the whole track. The settings for gain (dB) and frequency (Hz) can be adjusted for the three bands numerically as well as being used to open a graphical representation of the EQ.

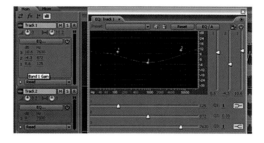

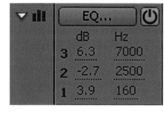

To adjust the EQ:

1   Highlight the EQ button ⇄ *fx* ⊩ ▥ at the top of the Main panel

2   Make adjustments numerically with the mouse by clicking on one of the three bands

3   Click the EQ button to open the graphical view to make your adjustments

> **Note:** Don't forget to turn on the power button for EQ ⏻

Track Automation Lanes

Located on each Audio Track of the main panel are Automation lanes showing parameter lanes for volume, pan, EQ and any Effects inserted on the track. These track automation lanes allow you to draw in setting changes with the mouse from the main panel or by using the slider controls within the mixer panel. Third party external mixer controllers (such as the Mackie control) can be used to input parameter changes. To show or hide the automation lanes, click on the small triangular arrow to the left of the automation status indicator at the bottom of the track control window. Multiple automation lanes are available per track by selecting the Show Additional Automation lanes [+] and choosing the appropriate parameter for that lane.

Clip Envelopes

As well as specific track automation lanes, Volume and Pan Clip envelopes can be automated on an audio clip basis. Located on the audio clip are Volume and Pan lines that you can draw in automation points.

This can be very useful for drawing complex Volume and Pan changes on a single audio clip. As seen in the diagram below you have the option of straight linear graph points and spline-based curves between the control points.

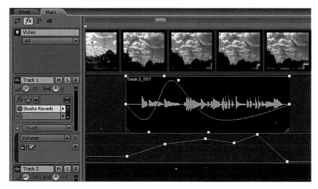

Volume and Pan Clip Envelopes and Automation Lanes

Automating Parameters/Automation Lanes

To draw in points manually:

1   Click on the show/hide automation lanes (triangular arrow at bottom of track)

2   Set the parameter type to appropriate title (e.g. volume, pan, effect)

3   Draw in automation points manually within the main panel using the mouse within the automation lane or by using the mixer panel slider controls to input changes

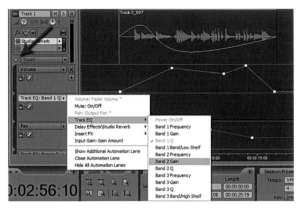

Multiple Automation lanes

To use the Mixer panel Automation:

1   Open the Mixer panel

2   Set automation mode to write

3   Altering any controls at all in the mixer will cause the changes you make to be written to the track

It's important to note that if you've added an effect to a track, like the Studio Reverb shown below, and you then move the controls within its control panel these are also stored as automation changes when you are in Write mode. You can open any track effect you've set up in the mixer panel simply by double-clicking on it.

Automation modes

- Off – no automation is read during playback

- Read – plays any automation applied to track during playback and mix down

- Latch – records automation as soon as you make an adjustment, overrides previous automation until playback stops

- Touch – only records when a parameter is touched. Reverts to previous recorded automation when you stop adjustments

- Write – records automation straightaway from playback, not waiting for any parameter adjustment

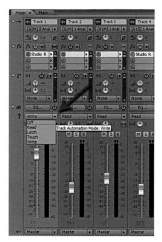

Automation modes within mixer panel

Automating Parameters/Clip Envelopes

To manually draw in Clip volume/Pan within the audio clip:

1 Locate to the position you want to add volume or pan automation

2 Click once to enter a volume point

3 Drag the points to raise or lower the volume

4 Add additional automation points along the audio clip as desired

To change the curves from linear to spline-based curves either:

1 Right click over a volume or pan point and choose Clip>Clip Envelopes>Pan or Volume>Use Splines

2 Choose Clip Envelopes>Volume>Use Splines

To crossfade between two audio clips:

1 Have the two clips to be crossfaded on separate tracks (sequential)

2 Also have the two clips that will become the actual crossfade section overlapping

71

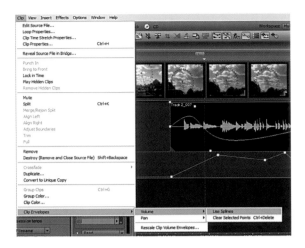

3 Using the Selection Tool select a range to become the crossfade, ensure both clips are highlighted by dragging the selection tool across both clips or with the CTRL key

4 Go to Clip>Crossfade

5 Choose one of the fade curves

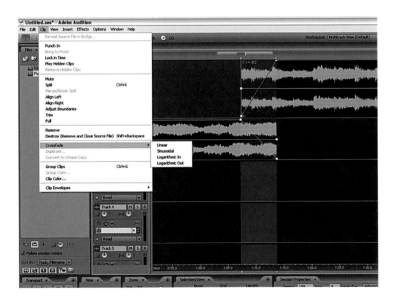

Types of Crossfades Curves

- Linear: straight line fade

- Sinusoidal: equal curve

- Logarithmic in: one clip fades in quicker

- Logarithmic out: one clip fades out quicker

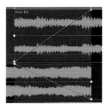 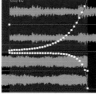 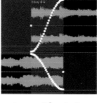 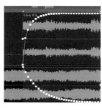

| Linear | Sinusoidal | Logarithmic in | Logarithmic out |

Using the Mixer

As Audition 2.0 uses ASIO compatible soundcard drivers, the whole mixing experience is faster and more responsive. Effects can be inserted and mixed in real time in tracks as well as being available for use in input monitoring. Audition includes several native VST plug-ins and supports hundreds of third-party VST plug-ins as well. All of this adds up to a real-world studio quality automated mixing environment on your personal computer. The mixer mimics the amount of tracks shown on your Main panel consisting of audio tracks and a single master output track. Additional bus tracks can be added and each track has its own volume, EQ and FX controls.

> **Note:** To open the mixer choose window>Mixer. You may want to dock the mixer next to the Main View for convenience.

Audition 2.0 revamped Mixer panel

The Channel Strip

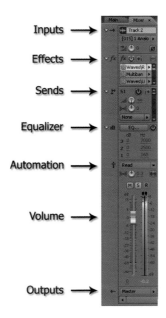

Inputs ⟶

Effects ⟶

Sends ⟶

Equalizer ⟶

Automation ⟶

Volume ⟶

Outputs ⟶

Input select

This is where you select the appropriate input from your soundcard for recording purposes. There is control for incoming signal level; signal phase can be set with the button next to input level control.

Effects

The Effects insert section lets you add up to 16 effects per track, options for pre- or post-fader and freeze are all located within this section.

- To insert an effect click on the insert FX Box, select an effect from the drop down menu

- To adjust the settings of the inserted effects double click the FX name to open the effect

- Freeze allows you to process the tracks effect onto the Audio clip, freeing up your computer processor. Simply click on the freeze button to process the effect

Sends

You can send as many tracks as you want to a Bus, where an effect, or several effects, can be applied to all of them simultaneously.

Bus setup Example

1   Set up a new Bus A and insert a reverb and compressor into its effects insert slots (if there is no bus, click add bus from the output Bus drop down box)

2   Set the outputs of the drum kit's separate drum tracks including Hi Hat, Cymbals, Snare, Bass drum etc. to this new Bus A track

3   Set the output of the New Bus A track to Master output

Now that the drums have been sent to the new Bus track all drums will have reverb and compressor on, and there will be one overall Bus volume fader for all the drums. On individual tracks, Sends to the Bus can be set to pre- or post-Fader, which means that they can be taken before or after the track EQ. Bus Send level and Pan control are also within this section.

Bus setup example showing all drum tracks sent to Bus A

Effects

Within this section you have a three band EQ

- To adjust the EQ scrub your mouse over the gain or frequency controls

- Click on the EQ button to open the full track EQ panel where you can save your presets

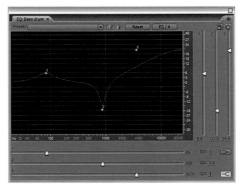

Equaliser Panel

Level Control

The Level Control allows you to adjust the Volume, Pan, Mute, Solo, Sum to mono and set record ready. There are also automation mode controls and the track output assign button here.

- To pan from L-R scrub over the pan knob

- To sum the track to mono click the Sum to mono button next to the pan knob

- To mute or solo select the M or S button

- To change the level simply raise or lower the volume slider

- Set the output either to a Bus track or a master track from your soundcard

CHAPTER 6
LOOPING CONTENT

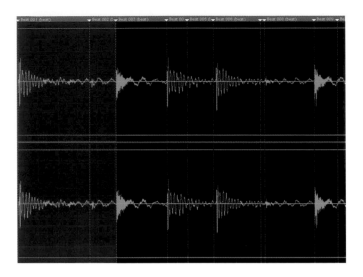

Introduction to Looping

Audition has the ability to find and calculate the beat and tempo of any track allowing you to splice up and loop sections of audio to create new songs. Looped audio can be heard in many different forms of music, including Rock, Jazz, Hip Hop, R&B and contemporary to name a few. As well as being able to start from scratch and building your own loops, Audition 2.0 comes complete with over 5000 royalty free Loopology drum beats, guitar, bass and synth parts already broken up ready for you to use and manipulate within your own project. Also, included are musical Beds and Sessions to help get you started.

> **Note:** It is illegal to use any loop or section of audio from someone else's record or song without permission from the owner.

In this section we will cover:

- Working with loops
- Creating loops

- Find beats and mark

- Define the loop range

- Calculate and adjust tempo

- Tempo match settings

- Using the included Loopology content

Working with Loops

To begin with, let's look at how a standard audio drum file can be quickly turned into a loop within the Multitrack View. Firstly, you will need to import an audio file into the Multitrack View. Notice that playing the file will result in the file playing only once and that the end of the file cannot be extended, unless of course the file has already been converted to loop.

To turn an Audio file into a loopable piece of audio:

1 With the file highlighted in the Multitrack View go to Clip>loop properties

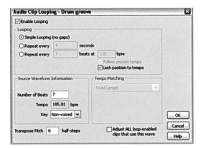

2 Enable the looping box

3 Single looping

4 Press OK

You have now converted the Audio file into a loopable piece of audio. Grab the corner of the audio and stretch it out, repeating (looping) the file as many times as you like. If you need to alter the tempo of a loop, then selecting the last of the three looping options will let you alter the speed in bpm that the loop plays using the method selected in Tempo Matching.

> **Note:** You should now have a Loop logo at bottom left of the audio clip showing that you are in loop mode.

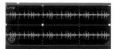

Find Beats and Mark (Edit View)

To get started let's look at how to create a new loop from an existing file and then to fine tune the start and end points ready for looping. To grab just the first bar of this two bar audio file we first need to locate where the beats actually fall so we know where to cut. Audition has a very handy tool that calculates and marks where these points are.

To calculate where the beats fall within the loop:

For this exercise you will need an audio file to work with (e.g. A drum loop)

1 Whilst in the Edit View highlight the audio file

2 Go to>Edit>Auto Mark>Find Beats and mark

3 The default settings for decibel rise is 10 db and the rise time is 9 milliseconds. These default settings can be customised to the level of the particular audio file you are working with, and determine the precise points at which the markers will be positioned. You can check the levels of your waveform by referring to the dB scale on the right hand side of the Main window, which will indicate the amplitude of different parts of the file

4 Press OK

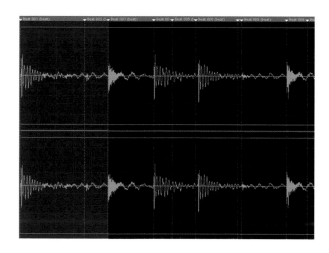

Result

Audition has calculated a number of beat marker points across the length of the waveform display placing them at the start of each hit point. Now that the hit points have been decided we can define a loop.

Define a Loop

To define a loop by deleting a section of audio at the start or end:

1   Set Snapping to Coarse and, using the beat markers points as snap points, select the audio range that you want for your loop with the Selection Tool

2   Trim (CTRL+T) the highlighted section, leaving just your loop material

3   Play the track and you should have a perfect loopable section

4   Go to File>save as

A new file has now been created which can be looped in the Multitrack View. If you turn on the metronome you may find that it is not in sync, so we will need to calculate the tempo and apply it to the audio file.

Calculate and Adjust Tempo

To set the file to loop and follow the session tempo:

1   Right click on the audio>File info (CTRL+P) Loop Info

2   Check that the number of beats is set to your original desired length (e.g. 4 beats is equal to 1 bars). If Audition has not set this exactly you may need to set the number of beats yourself. A quick way of checking is to play the track and count how many major beats there are from start to finish

3   Then you need only to set the Tempo Matching method and press OK

4   Import the adjusted file into the Multitrack View and it should now be in time with the project tempo

5   Adjust the tempo and see how the audio now follows in sync (turn on Metronome)

Note: You should now have a Loop logo at the bottom left of the file showing that you are in loop mode.

Tempo Matching Settings

- Fixed Length – file is unaffected in terms of tempo and pitch

- Time-scale Stretch – stretches the file without affecting pitch, similar to time stretch in the Multitrack View. It is best used for strings/pad sounds for best reproduction without artefacts

- Resample – file is affected in terms of pitch, similar to slowing down or speeding up a record player to match the tempo

- Beat Splice – best used to speed up drum or percussion loops and identifying each hit point to define tempo changes. You have the option to use files beat markers or to auto find beats

- Hybrid – uses the current Time-scale Stretch settings when slowing down and beat slicing when increasing the speed

Using the Included Loopology Content

Audition 2.0 now includes over 5700 royalty free musical loops, bed tracks, and sessions to get you started in creating your own music. Styles include rock, pop, classical, funk, lounge, house, techno, etc.

Loopology Song Creation Example

To locate the included Loopology content folder:

1 Locate to the Adobe Audition 2.0 folder on your hard-drive

2 Choose Content

3   Choose Loopology content

4   Import countrybrushdrum001.wav groove from loopology folder into files panel

5   Drop into Multitrack View

6   Drag out countrybrushdrum001.wav clip edge to repeat as many time as desired

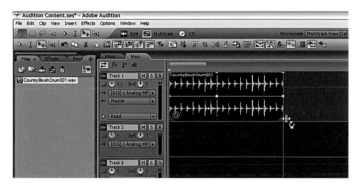

Single countrybrushdrum001.wav loop dragged out to repeat section

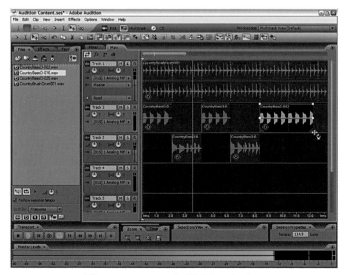

File arrangement in Multitrack view

7   Import Bass files from within the Country_bass-D_121BPM folder

8   Arrange in Multitrack View as desired

Experiment by changing tempo and adding in more loopology content to create your own unique songs.

CHAPTER 7
RESTORATION TOOLS

Overview

Probably the most amazing features of Adobe Audition 2.0 are the Restoration Tools. Live-recorded audio nearly always needs some kind of fixing, whether it be removing background noise from a film location recording, wind, hiss, camera noise, pops, clicks or taking out specific sounds. Adobe Audition 2.0 comes with several tools to get rid of unwanted sound. In this chapter we will go through the Restoration Tools step by step demonstrating methods of fixing your audio. The key feature Audition has is the Spectral Views, which allow you to see the audio in a whole new dimension. The Spectral views include:

Spectral Frequency Display

Shows the waveform in a frequency view. Bright colours showing high amplitude, dark colours showing low amplitude.

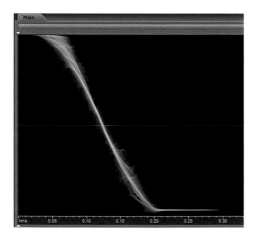

Spectral Pan Display

Shows pan information, top of display showing left channel, bottom of display showing right channel.

Spectral Phase Display

Shows waveform phase, in-phase being closest to centre and Out of Phase being closer to 180 degrees.

Spectral Analysis Tools

The Phase Analysis tool graphically shows if your audio file is Out of Phase. The graph shows phase information vertically as: In-phase in the top section and Out of Phase below the horizontal L and R line. The centre green ball represents In-phase material and turns red if the audio file is out of phase. The stereo image is shown along the horizontal line with mono information in the centre. Phase information can be adjusted using the phase invert button located on the input section of the mixer. The stereo image can be adjusted by using the Stereo Field Rotate and Stereo Expander effects under Amplitude.

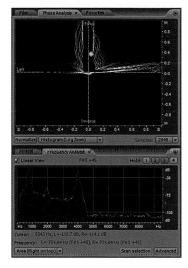

The Frequency Analysis illustrates a graphical view of frequency shown horizontally and amplitude shown vertically. Any frequency peaks that stick out will be shown in the graph allowing you to then lower or raise them using the EQ filter section. Hold points allow frequencies to be stored so you can see exact positional information.

> **Note:** Use the Zoom tool to zoom in to specific areas.

Remove Noise Procedure

If you have ever recorded live sound on a film shoot you will understand that background noise can be very annoying and destructive to that 'perfect take'.

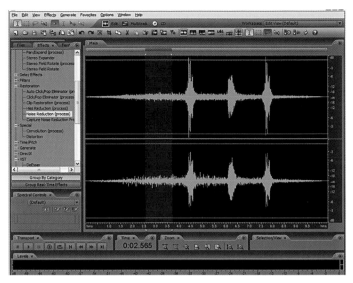

Standard View

As well as background noise, you can have camera noise, humming from neon lights, birds singing, dogs barking, transport noises, etc. Adobe Audition has some fantastic tools for removing background noise and specific unwanted annoying sounds.

The diagram above shows three main audio peaks (bird calls) surrounded by substantial background tube noise (Blue region). We want to remove the tube/background noise without affecting the three bird calls.

To remove background noise do the following:

1 Highlight a selection of noise that you want to delete with either the selection or Marquee Tool if you are in the Spectral Frequency display (you may want to zoom in by right clicking on the vertical dB display located far right)

2 Go to Effects tab>Noise Reduction then double click on Capture Noise Reduction Tool to capture the noise profile (or press ALT+N)

3 Double click Noise Reduction in the effects list

4   Preview and fine tune the noise reduction settings

5   Select entire file then press OK to process clip

After the remove noise effect has been processed, your end result will look something like the next graph. The background noise is totally removed while keeping our remaining audio perfectly intact.

Final Audio with background noise removed

Noise Reduction plug-in overview

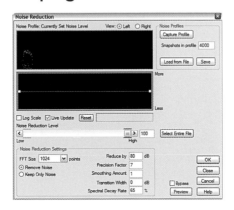

Noise Reduction procedure

The noise reduction plug-in is used to remove noise from any audio source. There are a number of factors involved in determine how much noise can be removed from a particular audio file. There are a number of settings that can be fine-tuned within the Noise Reduction plug-in and therefore a little trial and error is needed to get the best result.

Firstly, make sure you get the best possible sound profile area when Capturing a Noise Profile (ALT+N) as this will greatly determine what part of the sound frequency is being removed.

1 A graphical representation of the captured noise floor profile will appear in the top Noise Floor area

2 Select the entire file and press preview to listen to the file

3 Adjust the Noise Reduction slider to where the unwanted noise is no longer audible

4 Press OK

If you still want to make further reductions after this point try the following options:

5 Increase the Reduction graph by drawing in a boost area at the noise floor point

6 Adjust the Reduce by dB setting (increase/decrease)

Click/Pop Eliminator

Annoying Pops and clicks coming from old vinyl records, microphone pops and location recordings can be removed using the Click Pop Eliminator within Audition. As seen in the next diagram, substantial clicks and pops can be clearly seen throughout the clip.

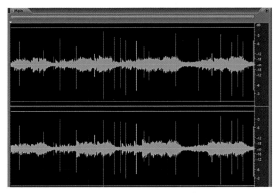

Showing multiple clicks and pops

To remove pops and clicks do the following:

1 Open an audio file into Edit View

2 Go to Effects tab>Noise Reduction

3 Double click on Auto Click/Pop Eliminator

4 Select the desired setting then press OK to process clip

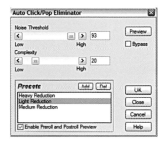

Note: For more advanced tweaking the click/pop eliminator under Effects>Noise Reduction>click/pop eliminator can be used, but the Auto Eliminator version does a great job.

Repair Transients

The Repair Transients tool is very similar to the clone tool within Adobe Photoshop. It allows you to replace your selection with areas of sound around it. The Audio file below is a suburban background atmos track with 2 X car horns peaks clearly seen in the Spectral views. We would like to remove the first car horn from the clip leaving the second car horn and suburban atmos track intact.

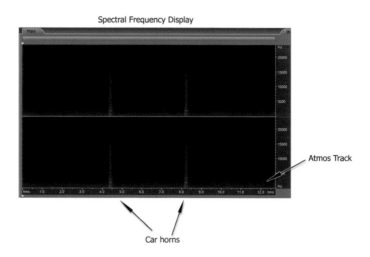

Spectral Frequency Display

Atmos Track

Car horns

Using the Repair Transients Tool

To remove specific sounds using the Repair Transients tool do the following:

1   Click on the Spectral View tab (or go to View>Spectral View)

2   With the Selection or Marquee Tool highlight the area of sound you want to remove

3   Choose Favourites>Repair Transients

4   Press OK

Repair transients on first Car Horn

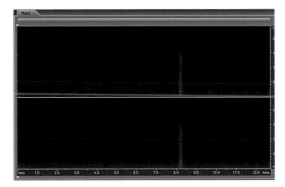

First Car Horn Removed using repair transients

Lasso Tool

The Audio clip example below is a suburban background atmos
track with two other louder sounds (a door bell and a car horn at the end).
Within Standard Waveform View we can clearly see the both doorbell and car
horn peaks (there is also some annoying camera noise which we cannot see from
this view).

The file shown within Spectral Frequency display visually reveals a doorbell,
car horn and camera noise tone as well as overall suburban atmos tone.

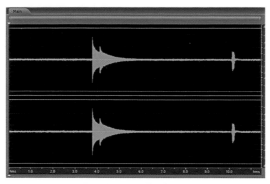

Diagram Standard View

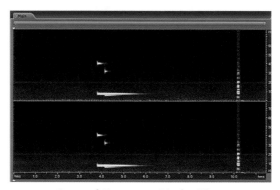

Spectral Frequency Display View

Removing specific sounds within Spectral View

To remove specific tones within the spectral Frequency view:

1 Open file into Edit View

2 Choose View>Spectral Frequency Display

3 Choose the Lasso tool and draw around specific sound

4 Press Delete

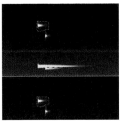
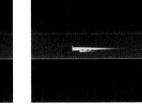

Lasso tool around clip   **Clip removed**

Marquee Tool

To remove specific sounds using the Marquee Tool do the following:

1. In the Spectral View select a range using the Marquee Tool

2. Go to Effects tab>Noise Reduction>double click on Capture Noise Reduction profile (ALT+N)

3. Double click Noise Reduction in the effects list

4. Select entire file then press OK to process clip

> **Note:** Noise reduction level can be decreased if too much of the sound is being removed.

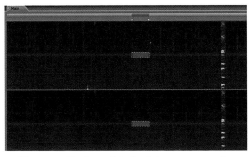

Marquee Tool used to isolate camera noise

97

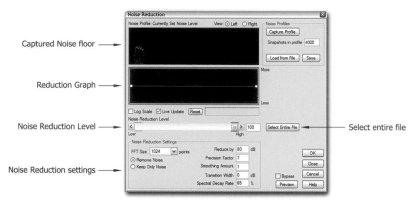

Captured Noise floor

Reduction Graph

Noise Reduction Level

Select entire file

Noise Reduction settings

Noise reduction dialogue box

Camera noise removed within Spectral view

Removing Vocal from Existing Mixes

In the past it has been impossible to go into an existing mastered music track and remove the vocal. You would need the original multi-track mix from the record company to then remix the track without it. Adobe Audition has a tool called a Centre Channel Extractor, which does exactly that (remove vocals) without affecting the stereo field. It can also be used to increase or decrease vocal levels on any stereo track.

To remove the vocal from an existing track do the following:

1 Load the track you would like to remove the vocals from

2 Go to Effects>Filters>Centre Channel Extraction

3   Highlight the desired effect from the Effect presets (remove vocal)

4   Select Male or Female voice under Frequency range

5   Preview the result then press OK

> **Note:** You may need to adjust or fine-tune the centre channel level and discrimination setting parameters to get the desired effect.

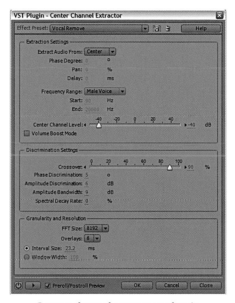

Center channel extractor plug in

CHAPTER 8
SURROUND SOUND

Surround Sound Overview

Surround Sound has been widely recognised in the film industry as a standard requirement and has been in use in differing surround formats since the 1950s. Today it has expanded into a widely used format for film, home cinema, DVDs and, more recently, digital TV broadcasting and the internet. Surround Sound, and in particular 5.1 Surround Sound, is becoming widely accepted and available as a listenable format.

The Surround Encoder

The Audition 2.0 Surround Encoder enables you to place a specific track in its own surround space. Often you will just want to place a specific sound in an area of the surround space and leave it there throughout the whole length of the track. Other times you will want to pan it around the surround space. Both are possible with the Adobe Audition Surround Encoder.

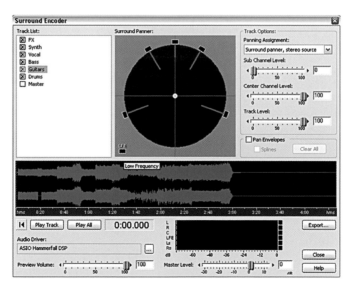

To open the Surround Encoder choose View>Surround Encoder from the Multitrack View. Once opened, all your tracks from the Multitrack View will be located on the Left hand side of the encoder under Track List.

Placing Tracks in the Surround Field

Highlighting a track will show where it is placed within the Surround Panner i.e. L+R, stereo.

To assign a track to a specific 5.1 output:

1 Select the track you wish to assign from the track list

2 Locate the appropriate output from the Panning Assignment drop down menu

3 Adjust overall track level, centre the sub woofer channels with the Centre and Sub Channel Level located under Track Options

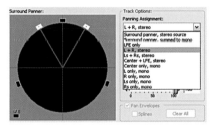
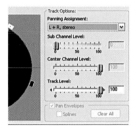

Panning Tracks in the Surround Field

At some point you will want to be able to pan a piece of audio around the surround field from front speakers to back speakers, etc. Adobe Audition enables you to draw in panning points along the length of the track.

To draw in surround points:

1 Highlight an audio track in the Track List, e.g. vocal

2 Drag the timeline cursor to the desired time point

3 Grab the white pointer within the Surround Panner Window to position the track

4   A new point will appear on the timeline below

5   To place a new surround point, locate a new timeline position and repeat steps 1 to 5

Previewing

Hardware Setup

To hear what's actually happening with your surround panning you will want to preview your work. This will require a soundcard with six separate outputs and you will also need a 5.1-surround monitoring speaker system.

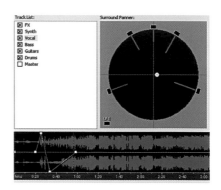

To set the hardware outputs:

1   Choose Edit>Audio Hardware Setup>Surround Encoder and map the separate 5.1 channels to the ASIO soundcard output ports

2   Press Apply>OK

To preview Surround Panning:

1   To hear a single track select the Play Track button from the Surround Encoder

2   To hear the whole mix select Play All button

> **Note:** There is a Preview Volume and Master Volume output level slider located at the bottom of the Surround Encoder Window.

Multi-channel Exporting

To export out a surround channel file:

1 Click the Export button on the Surround Encoder

2 Name the file under Multichannel Export Options session either:

- Export six individual mono files for use within another application

- Export as one interleaved, 6-channel wave file which is used to input into a DVD application such as Adobe Encore

- Export and encode as Windows Media Audio Pro 6-channel file, which is playable on a computer with Windows Media player or on the internet in surround sound. This is also a compressed format with options for Constant, Variable bit rates or Lossless settings

3 Finally press OK

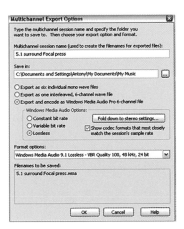

CHAPTER 9
WORKING WITH VIDEO

Importing Video Files

Adobe Audition 2.0 incorporates a new Media Player that can handle a larger range of video file formats. Along with AVI files Audition 2.0 now supports QuickTime, mpeg, mpg, mp2 and windows .wma formats.

> **Note:** You will need QuickTime installed on your computer before you can import QT files into Audition. Go to www.apple.com to download.

Having the ability to import video into your Multitrack View enables you to add sound FX, music and voice over to specific time code points along the timeline using the video as a guide.

To Import a video file:

1   From the Multitrack View go to File>Import Video File

2   Locate the Video File from your hard drive

3   Press Import

> **Note:** Only one Video File (track) can be loaded at a time. Once imported into your Project Folder, simply drag the file into the Multitrack Main View onto its own video track. You may wish to set the display format to SMPTE time code instead of Bars and Beats for editing sound to picture purposes.

Video Thumbnail Display Options

There are a few different Display Modes for viewing videos in the Multitrack View. You can choose all thumbnails, no thumbnails or just the first frame thumbnail within the clip.

To change the display settings:

1   Right click the Video Track and go to Thumbnail Options

2   Choose No Thumbnails, First Only or All

Time Stretching Audio To Fit Video

Once the video has been imported into the Main View you can move it around exactly the same as you would move audio. Often when working with video you will want to shorten or lengthen audio to fit tighter with the video track. An example would be the director wanting to re-cut or edit the video after you have already fitted your voice-over audio track perfectly, which means that the track will no longer work with the video. Without Time Stretching, this would mean recording the vocal take again.

Audition has a very handy tool called the Clip Time Stretching Toggle tool, which can speed up or slow down an audio clip to match the video scene changes without affecting the pitch of the audio. The tool looks similar to the trim/drag edge tool except it has a clock on the lower right hand corner.

Time-stretch tool

Time stretch tool showing stretch percentage amount

To Time-stretch an audio file to match the video track:

1 Choose the Clip time stretch toggle tool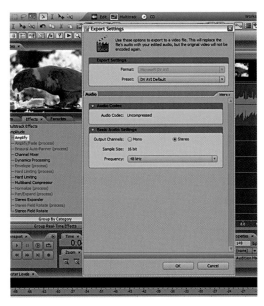

2 Click and drag the end of the audio clip left or right to time-stretch the audio faster or slower to fit the picture (the small stop watch appears on the trim tool)

3 Trim left or right to shorten/lengthen the audio file.

> **Note:** You can also access the time-stretch tool by holding down the CRTL key when placing the mouse at the edge of a clip (Same procedure as trim/drag edge tool).

Export Video

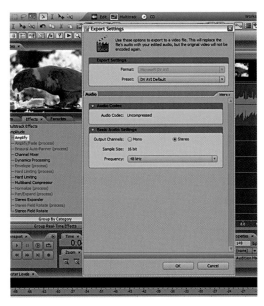

In the previous chapter we looked at how to import video with its accompanying audio into Audition 2.0, allowing us to enhance and restore the original audio file as well as being able to add additional dialogue, music and sound effects within the Multitrack View. At this point we now want to save the video, and replace the original audio with the new version with all the new changes and enhancements we have made.

To export all audio tracks within the Multitrack View into a single Video file such as a .mov or .avi file:

1 From the Multitrack view

2 Choose File>Export Video

3 Set export options

4 Press OK

5 Then choose a new name and location under Save as

Integration With Other Adobe Products

Adobe Audition 2.0 has been designed to integrate with the other Adobe Production Studio products. Whether you are using Adobe Premier Pro 2.0, After FX 7 or Encore 2.0, audio files can be edited, enhanced and mixed using Adobe Audition by using the Edit in Adobe Audition functions or Edit Original.

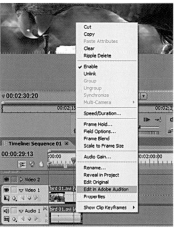

Premier Pro 2.0

To import a file into Audition from other Adobe Production Studio products:

1 Select the audio file that you wish to open from within the Adobe Production studio application such as Premier Pro 2.0

2 Choose Edit>Edit in Adobe Audition or right mouse click the audio File>Edit in Adobe Audition

3 Adobe Audition will launch with the file opening ready for editing

4 Once opened in Audition you can do any necessary editing or processing to the audio file

5 Finally Choose File>Save within Audition

111

6 Return to the Adobe Production studio application (i.e. Premier Pro) and the audio file will be automatically updated within its timeline or files panel

Note: Edit Original opens the file in whatever it was originally created in, whereas Edit in Adobe Audition opens a copy of the file directly within the Audition workspace.

CHAPTER 10
MASTERING, FINALISING

Overview of the Mastering Process

The mastering stage is the last stage of the recording process where all final adjustments happen. Audition has a number of quality tools to help you master your own projects on your home PC. In this chapter we will look at how to enhance the audio with a few of the included plug-in tools such as:

- Multiband Compressor – to compress individual frequency bands

- Parametric Equaliser – to achieve a balanced frequency response

- Hard Limiter – to increase overall output to be as loud as possible without distortion

Note: Mastering is a real art form and professional mastering engineers have years of experience, compared to a novice.

Dynamic Processing (Compression/Limiting)

A compressor reduces the dynamic range, which in a way is like squashing the audio and making a punchier, fatter sound. The level can then be increased to 0 dB thus creating a louder overall level. A Limiter can be used to raise the overall level of audio without changing or affecting the sound. If you are new to using the compressor try experimenting with the compressor presets to help you get started.

Multiband Compressor

The Multiband Compressor allows you to compress four frequency bands independently. Being able to allocate a separate compressor for the bass, low mid range, high mid range and high frequencies, allows you much greater control than just a single compressor in fine tuning the overall compressed output sound would.

- The crossover points are used to allocate a frequency range for each band

- Crossover points can be moved L or R to increase/ decrease frequency range for each band by either dragging in the main frequency band

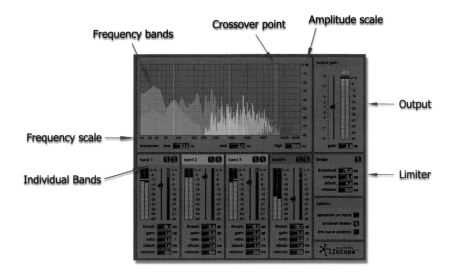

Frequency bands   Crossover point   Amplitude scale

Output

Frequency scale

Individual Bands

Limiter

window or by typing in values within the three crossover band dialogue boxes

- The frequencies are show in a visual display showing amplitude and frequency – different colours relate to the four individual band sections

- Four individual band controls for input level, compression ratio, attack & release, and gain

- Solo on each band allows you to hear frequencies within the separate bands

- The threshold level sets where the compressor starts to compress the signal

- The Gain adjusts the level after compression

- Attack and release settings for how fast/slow the compressor takes to start compressing

- There is an overall output gain and Limiter section

Solo/Bypass

Threshold level slider

Gain reduction meter

Input level meter

- The Margin level control located within the Limiter section allows you to set a final master output level so the signal level will never overload above 0 dB

Quick Multiband compressor setup:

1   Load the Multiband Compressor into the mastering rack or Effects Rack

2   Play the audio track and set the threshold level sliders of each band to the peak level of the Input Level Meter

3   Set the ratio on each band to around 1.5

4   Set Attack on each band to 10.0 ms and release to 100 ms

5   In the Limiter section set the threshold to –10 margin to –0.1

6   In the options panel turn on the Link Band controls tick box

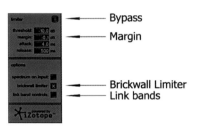 ——— Bypass
——— Margin

——— Brickwall Limiter
——— Link bands

Limiter sections

7   Play the track and slowly bring down the threshold level slider of all 4 bands (which are now linked) to increase the compression

Additionally, you can use the Compressor presets from the drop down menu to help you get started and understand the way the compressor works.

> **Note:** The Margin setting is the key to setting the peak output signal level to just below 0 dB.

Parametric Equalizer

The Parametric equalizer allows you to manipulate the tonal aspects of your track, giving you total control of the frequency spectrum. Included within the Parametric EQ is:

- A five band EQ each with separate frequency, Q, Gain and bypass controls
- Low shelf/high shelf cut-off

- Constant Q/width
- Second order buttons
- Ultra quiet mode
- Master Gain control

With the Parametric EQ it's easy to gently boost the higher frequencies using a board Q or removing troublesome frequencies using a narrow band Q/width setting. You can adjust settings for Frequency and amplitude by either using the sliders provided or by manipulating the frequency band points located within the main graph display.

The Parametric EQ gives much more flexibility and precision when compared to a standard graphic equaliser, which is limited by only having a fixed number of frequency bands and Q values to choose from.

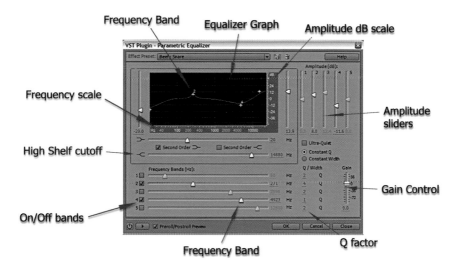

Equalizer settings

Frequency bands

Up to five Frequency bands can be switched on/off within the Parametric EQ, located to the left of the Frequency sliders. Turning on a band will reveal a green pointer within the graphical display representing its amplitude and frequency value.

117

To adjust the frequency and amplitude for each band

Use the five sliders located below and to the right of the main Graphical display or move the green pointer within the graphical display. Raising the pointer increases amplitude. Lowering pointer decrease amplitude. Moving pointer L or R sets the band frequency.

Q/Width

The Q/Width represents the width of the selected frequency band. A narrow Q can be selected to pin point frequencies to be reduced or increased and a wide Q can be selected to increase or decrease a broader range of frequencies.

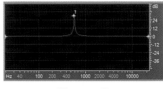

Narrow Q

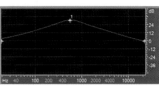

Wide Q

To set the Q/width:

1 Turn on a frequency band

2 Set the Q/width amount located next to each frequency slider

To find a troublesome frequencies to be removed:

1 Turn on a frequency band

2 Set the Q/width to a narrow setting (50–100)

3 Increase its amplitude to above 0dB, and sweep the frequency control through the frequency spectrum whilst listening to find the troublesome frequency

4 Once found, decrease amplitude until it is no longer audible

Low Shelf and High Shelf Cutoff and Level

Use the low shelf cutoff and level to reduce low bass rumble. Use the High shelf cutoff and level sliders to either decrease unwanted hiss or boost the high treble frequencies.

The diagram below shows Low shelf cutoff frequency and level set to reduce low-end rumble by −30.5 dB at 28 Hz and High shelf cutoff frequency and level set to boost treble by 19.6 dB at 11182 Hz.

Low Shelf Level   High Shelf Level

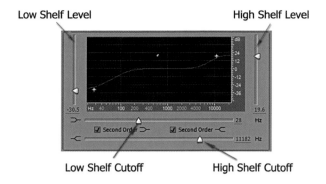

Low Shelf Cutoff   High Shelf Cutoff

Second Order Button

Adjusts the low and high filter shelves to have a slope of 12dB per octave, instead of the default first order 6dB per octave.

Constant Q/Width

Describes the frequency band width as either Q value or absolute width value in Hz.

Ultra Quiet

Eliminates any noise or artefacts but uses significantly more processing power.

Gain Control

Allows you to compensate the output level after you have made EQ adjustments.

Hard Limiter Overview

The Hard Limiter allows you to greatly increase the overall volume by pushing the input signal and limiting the output. By boosting the input, the signal will limit at the point set by the 'Limit Max Amplitude to' slider, creating an overall impression of increased loudness.

Limit Max Amplitude – sets the final overall output level

Boost Input By – increases input level volume before it hits the output hard limiter

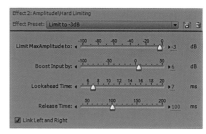

Hard Limiter

Look-ahead Time – the time that the limiter is able to pre-judge the signal level for, and react to it

Release Time – the time it takes for the limiter to drop back 12 dB (roughly the time it takes to recover from an extremely loud peak)

Link L & R – links left & right signal

Quick Start

1. Set the desired amplitude output level close to 0 dB (−.2 dB)

2. Increase the boost input fader

3. Check Audition's master volume level (which should be exactly the same as the Limit Max Amplitude setting)

4. Experiment by trying the effects presets from the drop down menu

5. Save your own presets by choosing the Add Preset button next to the bin

CHAPTER 11
EXPORTING, SAVING FILES/PROJECTS

Export Audio (Multitrack View)

After completing the mixing process you will need to export all the separate audio tracks and parts from your multitrack session to a final stereo or mono file ready for distribution. By default all tracks within your multitrack session are sent to the Master output bus. Each track within your session can be assigned to an output from either the Main Multitrack View or the mixer as shown below.

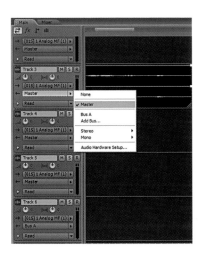

To export an audio file:

1 Within the Multitrack View choose File>Export Audio Mix down

2 The Export Audio Mix down dialogue box opens where you can specify a name, location and file type

Mix Down Options

On the right hand side of the Export Audio Mix Down dialogue box you have options to export from a specific source. If you are unsure, select Master as this is the main output bus. Output options are either:

1 Choose Master from the Source View to mix all tracks from your Multitrack session down to a final stereo or mono wave file (Default session output Bus)

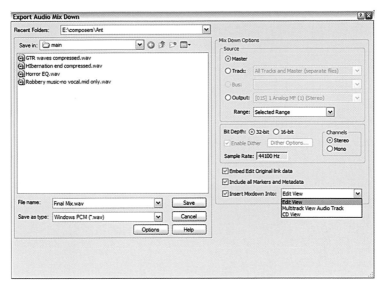

Export audio mix down dialogue box

2 Choose Track to export a single audio track from your multitrack session

3 Choose Output to export audio routed to a selected hardware output – the default is the output selected for the Master channel, but if you have other buses with independent output channels set on them, they are available here, so that you can save whatever has been sent to them

4 Set the Range to either entire session or range

5 Set the desired bit depth and channels

Note: You will need to have selected a range first within the Multitrack View before this option will pop up.

Other options include Embed Edit Original link data. This stores the Audition project information within the file so that Audition can be opened at a later point from within another Adobe application such as Premier Pro. Optional Insert mix down into Edit View, Multitrack View or CD tick box that allows your mixed down file to be inserted directly after export, e.g. into Edit View ready

for mastering. Bit Depth can be set to either Mono or Stereo, 16-bit for CD compatibility or 32-bit for archiving or using within another application such as Premier Pro or After FX to retain audio quality.

Audio File Formats (mp3, wav)

There are a number of audio file format options to export to from the 'Save as type' drop down dialogue box.

Export Audio (Edit View)

When in Edit View there are a number of ways to save your mono or stereo edited file. You have several exporting/saving options so we'll go through them one by one:

- Save – choosing save will save over your original file (be careful)
- Save As – saves the file under a new name or file type
- Save Copy as – to save a copy under a new name or file type keeping your original file within the project

- Save selection – allows you to save a selection only, great for saving a small section of audio within a large file. Use the range tool to first select a range and then press save selection

- Save All – saves all opened files within the project

- Revert to saved – reverts back to original saved file

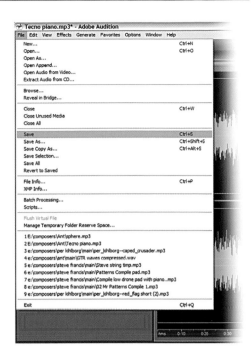

Batch Processing

The batch process function allows you to process multiple audio clips.

Batch processing function:

- Run a script – an example would be to normalise a batch of files

- Resample – convert sample rate

- Save files to new audio format – convert .wav to .Aiff

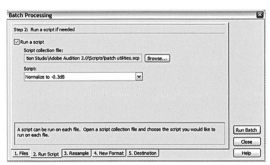

- To open the Batch Processing choose File>Batch Processing from within the Edit View

The batch processor runs in a sequence. If you run through the tabs at the bottom in turn, filling in the relevant sections, you will arrive at a point where the batch sequence can run, and the 'Run Batch' button becomes un-greyed.

CHAPTER 12
MAKING A CD

CD Project View

Audition comes complete with CD burning capabilities built in, allowing you to add tracks that you have mastered to a CD list. They can then be arranged into an appropriate order and burnt to CD all without needing any separate CD burning software. Whatever bit or sample rate you have been working with, Audition 2.0 will convert on the fly all files to the 16-bit/44.1 k stereo CD format, streamlining the CD burning process.

Equal volume can be set for all tracks within the CD View by utilising the Group Waveform Normalize procedure.

To add audio files to the CD project View:

1   Select CD View tab

2   Import files into the CD Files view

3   Highlight the files in the Files pane

4   Either click the Insert into CD List tab button located at the top of the Files View or simply drag the files into the Main CD area

To Change the Order of Files

Once the audio files are listed in the CD View you can re-arrange the order from first to last.

To change the stacking order of your audio files:

1 Highlight a file in the main CD list

2 Select the appropriate button on the right hand side to Move up or Move down within the CD list or click and drag the file to new position

To remove a file from the CD list:

1 Select the file

2 Choose remove or Remove all to remove the entire list

Properties Setting

Track properties such as title, artist, and copy protection can be set for each audio file within the CD view.

To locate the Track Properties window:

1 Highlight an audio file within the CD view

2 Click the Track Properties tab on the right

Group Waveform Normalize

By now you will have mixed and mastered all of the appropriate audio files and are now ready to burn a CD. If you are burning more than one audio file you will probably want all the levels of the tracks to be the same so that the listener will not need to adjust the volume control at any stage. Audition has included a Group Waveform Normalize function that sets a constant level for all tracks.

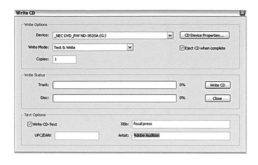

To Normalize a group of files:

1 Add files to the CD View by high-lighting the file and pressing the Insert into CD list

2 In the CD View choose Edit>Group Waveform Normalize

3 Highlight the files to be Normalized (choose All)

4 There is an option to analyse the files to observe the maximum output level

5 Run Normalize

Burning a CD

Finally to burn a CD:

1 Place a blank writeable CD in your drive

2 Click the Write CD button located at the bottom of the CD View or go to File>Write CD

3 Within the Write CD properties section select your CD drive from the device list

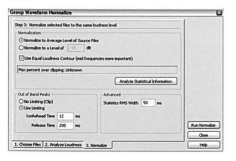

4 Select appropriate write mode

5 Select amount of copies

6 Enter in title and artist if desired by ticking the Write CD-text button

7 Press Write CD

Note: Not all CD burners support writable text.

Index